Mistral: The Complete Guide

Europe's Flagship AI — the Vibe Assistant, Open-Weight Models, and Data Sovereignty

Ken Tannenbaum

2026

Finnoybu Press

Mistral: The Complete Guide

2026 Edition

Europe's Flagship AI — the Vibe Assistant, Open-Weight Models, and Data Sovereignty

---

Published by Finnoybu Press

For permissions and inquiries: press@finnoybu.org

First Edition: 2026

About the Author

Ken Tannenbaum is a technologist, entrepreneur, and AI governance researcher. He is the founder of Finnoybu Press and the founder/steward of the AEGIS Initiative, an open framework for AI safety and accountability. Learn more at aegis-initiative.com.

When he's not building governance standards for artificial intelligence, he's helping everyday people understand and use it.

Dedication

For the teams who refused to accept that capable AI and control over your own data are things you must choose between.

The AI for Everyone Series

This book is part of the **AI for Everyone** series by Finnoybu Press, designed to help real people use AI tools effectively. Visit **press.finnoybu.org** for the latest titles and updates.

Collection One: Product Deep Dives

- *ChatGPT: The Definitive Guide (2026 Edition)*
- *ChatGPT: The Power User Guide (2026 Edition)*
- *Claude: The Definitive Guide (2026 Edition)*
- *Claude Code: The Power User Guide (2026 Edition)*
- *Gemini: The Complete Guide (2026 Edition)*
- *Gemini: The Advanced Guide (2026 Edition)*
- *Copilot: The Workplace AI Guide (2026 Edition)*
- *Copilot: The Advanced Workplace Guide (2026 Edition)*
- *Perplexity: The Research Playbook (2026 Edition)*
- *Perplexity: The Advanced Research Playbook (2026 Edition)*

Collection Two: Product Guides

- *NotebookLM: From Sources to Audio Overviews (2026 Edition)*
- *DeepSeek: Code, Math, and Reasoning (2026 Edition)*
- *Cursor: The AI Code Editor (2026 Edition)*
- *OpenClaw: Agentic AI for Everyday Work (2026 Edition)*

Collection Three: Cross-Platform Skills

- *AI Systems Playbook (2026 Edition)*
- *AI Prompt Engineering (2026 Edition)*
- *AI for Developers (2026 Edition)*
- *AI for Business (2026 Edition)*
- *AI for Students (2026 Edition)*
- *AI Governance for Practitioners (2026 Edition)*
- *Securing Agentic AI (2026 Edition)*

The Extended Series — Product Guides, Volume II

- *Codex: The Cloud Coding Agent (2026 Edition)*
- *Grok: Real-Time AI from xAI (2026 Edition)*
- *Mistral: Vibe and the Open European AI (2026 Edition)*
- *DeepL: Translation, Writing, and Voice (2026 Edition)*

The Extended Series — The Capability Set

- *Automate Your Work with AI (2026 Edition)*
- *Build Apps Without Code (2026 Edition)*
- *AI Literacy for the Workplace (2026 Edition)*

Table of Contents

Introduction

I wrote this book because the American AI giants ask you to make a choice that Mistral was built to refuse — and the people who feel that choice most sharply are exactly the ones the flashier tools tend to forget.

The choice is this: capability or control. For most of the short history of modern AI, the most capable assistants lived on a handful of US companies' servers, reachable only by sending your data to them, on terms you could not negotiate. For a great many people that is fine. But for hospitals, banks, law firms, government bodies, European companies under GDPR, and anyone with information they are not permitted to export, it is a wall. The capability is real and, for them, unreachable — not because the model is not good enough, but because they are not allowed to use it the only way it is offered.

Mistral is Europe's answer to that wall, and its bet is that you should not have to choose. By publishing its model weights under genuinely permissive licenses, offering European infrastructure, and supporting self-hosting and private deployment, Mistral turns "capability *or* control" into "capability *and* control." That conjunction — not a benchmark score — is the reason it exists as its own thing rather than a European echo of an American idea, and the reason it deserves a book of its own.

What This Book Covers, Briefly

Mistral is Europe's flagship AI company, and its work comes in two halves this book treats together: **Vibe**, the AI assistant — formerly Le Chat, renamed in 2026 as it grew from a chatbot into a work agent, organized around three modes (Work, Code, and Chat) — and a family of **open-weight models** you can download, inspect, run yourself, fine-tune, or reach through an API, built so your data can stay under your control.

This book covers all of it: getting started with Vibe and its three modes, using it for real work and for code, and then the heart of the matter — what "open weight" actually means, running the models yourself, data residency and sovereignty, and enterprise self-hosting. It covers building on Mistral through La Plateforme, the business and European-compliance case, and an honest comparison with the US flagships. And it closes, as this series always does, with where Mistral falls short and how to use it responsibly.

Why This Book Exists as Its Own Thing

There is an obvious objection to a whole book about Mistral, and I want to meet it head-on, because if you are holding this volume you have probably already half-formed it: *isn't this just ChatGPT with a French passport, and couldn't I learn it in an afternoon from the documentation?* The answer is no, and the reason is the same reason the book is organized the way it is. Every other major assistant asks you to learn one thing: how to talk to a box that lives on a company's servers. Mistral asks you to learn two things at once — how to use a capable assistant, and how to decide *where the capability runs and who controls the data while it does.* That second skill has no equivalent in the closed world, because in the closed world the answer is fixed: the model runs on the vendor's servers, the data goes there, and the terms are theirs. With Mistral the answer is a spectrum you choose along, and choosing well is most of using the tool well. A book that taught you only the chat box would have taught you the smaller half.

So the structure mirrors the two skills. The early parts teach the assistant — the modes, the conversations, the work and the code — the way any practical guide would. The middle parts teach the thing that has no analogue elsewhere: what it means to take a frontier model home, run it on your own hardware, keep your data on your own soil, and answer to your own regulators instead of a foreign vendor's terms of service. You can read either half on its own. But the book treats them together because Mistral does, and pretending otherwise would leave you with half a tool.

A Word About Where the Industry Stands

It helps to know the shape of the field this book sits inside, because Mistral does not exist in a vacuum and its choices only make sense against the backdrop of what everyone else chose. As of mid-2026, the frontier of AI is held by a small cluster of American labs and a comparable cluster of Chinese ones, almost all of which ship their most capable models as closed services — you reach them through an API or a chat app, the weights never leave the provider, and the relationship is rental, not ownership. That is the default the whole consumer and enterprise market grew up inside, and for most users most of the time it is genuinely fine. The closed model is convenient, it is maintained for you, and the vendor absorbs the operational burden.

Mistral is the most prominent counterexample at the frontier: a lab that ships genuinely capable models as downloadable weights under permissive licenses, deliberately, as its core strategy rather than as an afterthought or a research gesture. It is not alone in releasing open-weight models — there are others, and the open-weight world is healthier and more crowded than it was a year ago — but it is the most prominent one positioning openness and European sovereignty as the *whole point* rather than a side dish. Keeping that backdrop in mind makes the rest of the book land correctly: Mistral's distinctiveness is not that it is the smartest model in the room, because often it is not, but that it is the most capable model you are genuinely allowed to take home.

Who This Book Is For

- **Individuals and teams** who want a capable, privacy-respecting assistant aligned with European data values.
- **Developers and builders** who want open-weight models they can run, fine-tune, and deploy on their own terms — for cost, for control, for freedom from lock-in.
- **Regulated and sovereignty-conscious organizations** — European institutions, regulated industries, the public sector — for whom data residency and self-hosting are requirements, not preferences, and for whom Mistral is often the *only* option that clears the bar.

A few words on how those three groups will actually use the book, because they read it differently. The individual reaching for Vibe to draft emails and chew through documents wants Parts I through IV and can treat the sovereignty material as background — interesting, but not load-bearing for the way they work day to day. The developer wants Part VI most of all, and will dip back into Part V for the self-hosting mechanics; for that reader the assistant chapters are orientation, and the API is the product. The regulated organization is the reader for whom this book is, frankly, most necessary, because they are the ones for whom "which deployment, on whose infrastructure, under which license" is not a preference but a compliance question with legal weight — and for them Part V and Part VII are not optional reading but the core of the case they will have to make internally to whoever signs off on the tool.

You do not need to be technical for most of this book. The developer and self-hosting material in Parts V and VI is signposted so you can navigate around it; everything before assumes only curiosity and basic comfort with the tools you already use.

How to Read This Book

The book is in eight parts, plus reference.

Part I: Meet Mistral is orientation — what Mistral is, where it came from and why Europe wanted its own champion, how Le Chat became Vibe, and what the model family contains.

Part II: Getting Started with Vibe is the practical setup — the access plans, the three modes that define how you use Vibe, your first conversations, and the Libraries, Connectors, and Memory that make it persistent.

Part III: Vibe Work covers the agentic side — writing and documents, research and the web, and building agents in the no-code builder.

Part IV: Vibe Code covers software development — coding in Vibe and the CLI and editor integrations.

Part V: Open Weights and Sovereignty is the heart of the book — what "open weight" actually means, running the models yourself, data residency and sovereignty, and Vibe Enterprise.

Part VI: Building on Mistral is for developers — La Plateforme, choosing a model, fine-tuning, and the cost and deployment options.

Part VII: Mistral for Work is the business case — real use cases, the European compliance story, and where Mistral fits beside the US flagships.

Part VIII: Judgment and Limits is the honest close — what Mistral gets wrong, and responsible use.

Reference wraps with a model reference, an API quickstart, self-hosting notes, a glossary, and resources for staying current.

Read straight through or jump around. New to Mistral: Parts I and II. Here for the sovereignty story: Part V, then Part VII. Building on it: Part VI. Evaluating it against what you use: Chapter 24, then Part VIII. If you have exactly one hour and want the spine of the argument rather than the full tour, read Chapter 1, then Chapter 14 on what "open weight" actually means, then Chapter 24 on where Mistral fits beside the American flagships — those three together carry the thesis, and everything else is the working detail that fills it in.

A Note About Accuracy

This corner of the AI market moves in weeks, not years. Model names, prices, licenses, plan boundaries, and features will have drifted by the time you read this. This book is accurate as of mid-2026.

The durable parts — what open weights and sovereignty are for, how to choose a deployment, where to trust the tool and where not to — will hold. The model version numbers and the price points will not. Where a detail changes fast, I flag it and point you at Mistral's current documentation rather than pretending it is permanent. This is true with extra force for the open-weight licenses, which vary by model and have shifted before: read the actual license for the actual model, every time.

A small note on how to read those flags, because they are not throat-clearing. When I write "as of mid-2026" or "confirm against the current docs," I am not hedging to cover myself — I am telling you precisely which sentences to distrust once this book has aged a little. A claim about *why* open weights matter to a hospital will be as true in three years as it is today; a claim about which model carries which license, or what a million tokens costs, may be wrong by the time the ink is dry. Treat the flagged specifics as a snapshot and the unflagged principles as the lasting content, and the book will keep serving you long after the version numbers have turned over.

Let's Get Started

Mistral is not the most capable model on every benchmark, and this book will not pretend it is — Chapter 24 places it honestly beside the American flagships. What Mistral is, more than any of them, is *capable AI you are actually allowed to control* — the assistant the hospital and the bank and the ministry can run on their own terms, the open weights the builder can fine-tune without asking permission, the sovereign deployment that keeps the data home. For the people who need that, it is not one option among many. It is the only one that clears the wall.

Turn the page. Let's start with what Mistral is.

Part I: Meet Mistral

Audience: everyone.

"The question was never just 'how capable is the model.' For a lot of the world it was 'am I even allowed to use it' — and that is the question Mistral set out to answer."

— A working principle

Chapter 1: What Mistral Is

A few years ago I sat in a meeting where a hospital's technology lead explained, with real frustration, why her team could not use the AI assistant everyone kept telling her was revolutionary. The capability was not in question. The problem was that using it meant sending patient information to a company's servers in another country, governed by another country's laws, with data-handling terms she was not allowed to accept and could not negotiate. The tool was brilliant and, for her, unusable. She did not need a better model. She needed a model she was permitted to use — one whose data stayed where her regulators required, ideally one she could run on infrastructure she controlled. At the time, her options were thin.

That gap — between the most capable AI and the AI you are actually allowed to run on your own terms — is the gap Mistral was built to close, and it is the reason this book exists as its own thing rather than a footnote to the American assistants. Mistral's bet is that capability and control are not a trade-off you should have to make. This chapter is about what Mistral is, what it isn't, and why a company whose defining trait is *openness and sovereignty* turned out to matter most to exactly the people the flashier tools could not serve.

The One-Sentence Definition

Mistral is Europe's flagship AI company, and its work comes in two halves that this book treats together: **Vibe**, a frontier AI assistant for everyday and professional work — organized around three modes, Work, Code, and Chat — and a family of **open-weight models** you can download, inspect, run yourself, or reach through an API, released under genuinely permissive licenses and built so your data can stay under your control.

Several words in that sentence carry the weight.

"Europe's flagship AI company" — Mistral is French, founded in 2023 by researchers out of the labs behind some of the field's foundational work, and it has become the credible European answer to the American and Chinese giants. That is not a patriotic footnote. It shapes everything about the product: the regulatory posture, the data-residency story, the alignment with European values around privacy and sovereignty, and the willingness to give things away that the US labs keep locked up. When the rest of this book talks about sovereignty as a feature, this is where it comes from.

"Vibe" — the assistant most people will actually touch. Until 2026 it was called **Le Chat**; it was renamed Vibe as it grew from a chatbot into a work agent. Vibe is organized around three modes — **Work** for delegating multi-step tasks across your apps, **Code** for software development, and **Chat** for the classic conversational experience — and Chapter 6 is devoted to the distinction, because choosing the right mode is most of using Vibe well. If you have heard of "Le Chat," Vibe is what it became; Chapter 3 tells that story.

"Open-weight models" — this is the part the American flagships do not match, and the heart of what makes Mistral different. Mistral *publishes the weights* of its models — the actual trained parameters — under permissive licenses, so that you, or your company, can download them, examine them, run them on your own hardware, fine-tune them, and deploy them without asking anyone's permission or sending anyone your data. That is a fundamentally different relationship with an AI than renting access to a black box over someone else's API. Chapter 14 unpacks exactly what "open weight" does and does not mean, because the phrase is slippery and the details matter.

"So your data can stay under your control" — the practical payoff, and the hospital lead's whole problem solved. Because you can run the models yourself, or use Mistral's EU-based infrastructure, or deploy in a private cloud, you are not forced to hand your sensitive information to a foreign black box as the price of using capable AI. For regulated industries, European institutions, and anyone for whom data residency is not optional, that is the difference between "interesting" and "usable."

To make that abstract sentence concrete, picture the hospital lead's options laid side by side. With a closed American assistant, the only path is the API: a patient's record leaves her network, travels to servers in another jurisdiction, gets processed under terms she cannot edit, and she has to trust a contract and a foreign legal system that the data was handled as promised. With Mistral she has a menu instead of a mandate. She can run an open-weight model on a server inside her own hospital, in which case the patient record never crosses the building's firewall and there is nothing to trust because nothing left. She can use Mistral's European infrastructure, in which case the data is processed under

European law on European soil — still hosted, but hosted somewhere her regulator recognizes. Or she can stand up a private cloud deployment that sits between the two. The same underlying capability reaches her three different ways, and she picks the one her compliance officer will sign. That menu is what "control" means in practice, and it is the thing no amount of model capability can substitute for when the law says the data may not leave.

A Day With Mistral, Concretely

It helps to see the abstractions doing real work, so here are three short, deliberately ordinary scenes from people using Mistral well — none of them dramatic, all of them the kind of thing that happens a hundred times a day.

A solo consultant in Lyon opens Vibe in the morning, drops in a forty-page tender document a prospective client sent overnight, and asks for the three obligations that would be hardest for a small firm to meet and where in the document each one hides. Two minutes later she has a short brief, the page references to check, and a draft of the clarifying email she needs to send back. She is using the hosted assistant like anyone would use any assistant — the sovereignty story is, for her this morning, mostly invisible. It matters to her only in the loose sense that the company behind the tool is European and she feels easier about where the tender text goes. This is the most common way Mistral gets used, and it is worth saying plainly that for this user it competes on being a good assistant, full stop.

A developer at a mid-sized German software company has wired a smaller Mistral model into the company's internal support tool to triage incoming tickets — classifying them, drafting first-pass replies, flagging the ones that need a human. He chose Mistral not because it topped a leaderboard but because he could run the model on the company's own servers, which meant the customer messages flowing through the classifier never left the building and his security team approved the project in a single short meeting instead of a quarter-long vendor review. The capability mattered; the *deployability* is what got the project shipped.

A records officer at a regional government body uses a privately deployed Mistral instance to summarize and route citizen correspondence. For her the entire value proposition is the part the consultant never thinks about: the deployment was the precondition for the project existing at all, because routing citizens' personal data through a foreign cloud was simply not permitted, and a less capable model she was *allowed* to run beat every brilliant one she was forbidden to touch. Three users, three relationships to the same company — and the only thing that separates them is which half of Mistral they are leaning on.

What Mistral Is Not

Five clarifications up front, because each is a misconception I have watched send people the wrong way.

Mistral is not a US-flagship clone with a French accent. It would be easy to assume Mistral is just "ChatGPT, but European," and to evaluate it purely on whether its chat experience matches the Americans feature for feature. That misses the point. Mistral's reason to exist is the open-weight, sovereign, run-it-yourself model — a different value proposition, not a worse copy of the same one. Judge it on that, and the picture changes.

Mistral is not only Vibe, and Vibe is not only a chatbot. The assistant is one surface. Underneath sits the model family and **La Plateforme**, the API and developer platform (Chapter 18), plus the option to self-host (Chapter 15) and deploy privately (Chapter 17). Many of Mistral's most important users never open the Vibe chat box at all — they build on the models directly. And Vibe itself is now a work agent that delegates multi-step tasks, not just a question-answerer.

"Open weight" is not the same as "open source," and Mistral is not uniformly open. This is the single most important nuance in the book, and Chapter 14 is built around it. Mistral releases model *weights* under permissive licenses — but the licenses vary, some models are more open than others, and "open weight" is a narrower and more practical thing than the full "open source" ideal. Mistral's most capable model has at times been the most open; other times a model is held back. The honest frame is "meaningfully more open than the American flagships, with real and shifting limits," not "everything is free for everyone."

Mistral is not always the most capable model on the leaderboard. Mistral is a frontier-class lab and its top models are genuinely competitive — but it is a smaller company than the American giants, with less compute and a smaller ecosystem, and on raw benchmark-topping capability the US flagships often lead. Mistral's argument is not "we are always the smartest." It is "we are highly capable *and* open *and* sovereign," and that combination is worth more than a few benchmark points to the people who need it. Chapter 24 makes that comparison honestly.

Mistral's sovereignty story is strongest at the enterprise surface. If you are an individual using the free Vibe app to draft emails, the "data sovereignty" framing is mostly abstract — you are using a hosted assistant like any other, and the privacy terms are what matter (Chapter 16). The full power of the sovereignty proposition — self-hosting, data residency guarantees, private deployment — lives at the business and enterprise level (Part V). Knowing which half of Mistral you are using keeps your expectations honest.

Why Openness and Sovereignty Are the Whole Game

It is tempting to evaluate an AI company by whether its assistant tops the others on a clever benchmark. Mistral's real advantage is harder to put on a leaderboard and more decisive in practice: it gives you a capable model you are actually *allowed to control.*

Every closed, API-only assistant asks you to accept the same arrangement: the model lives on the provider's servers, your data goes to those servers to be processed, and the terms are the provider's to set. For a great many users that is perfectly fine. But for a large and important set — hospitals, banks, law firms, government bodies, European companies under GDPR, anyone with secrets they cannot export — that arrangement is a wall. They are not allowed to send the data, full stop, and no amount of model capability gets them over it.

Mistral's openness turns the wall into a door. Because the weights are published, the data does not have to leave. You can run the model inside your own network, on your own hardware, under your own rules, with nothing transmitted to anyone. Or you can use Mistral's European infrastructure and get data-residency guarantees a US-based provider structurally cannot match. The capability becomes *reachable* for people who were locked out — and that reachability, not a benchmark, is the whole game. A slightly less capable model you are permitted to use beats a slightly more capable one you are forbidden to touch, every single time.

That is why open weights and sovereignty get their own Part (Part V) and why this chapter keeps returning to them. They are not ideological flourishes. They are the mechanism by which Mistral becomes usable for the people who need it most, and the reason the company is more than a European echo of an American idea.

There is a second-order benefit here that is easy to miss and worth naming, because it changes how you should think about commitment to the tool. When you build on a closed, API-only assistant, you are renting a capability that the provider can reprice, restrict, deprecate, or discontinue on a timeline you do not control — and people who built businesses on top of model APIs have, more than once, woken up to find the model they depended on retired or the terms rewritten. An open-weight model you have downloaded does not stop working because a company changed its mind. The weights you have are the weights you have; a license already granted on a release already shipped cannot be retroactively clawed back from the copy on your disk. That permanence is its own kind of sovereignty — sovereignty against the vendor's future as much as against any government's borders. For an institution making a multi-year bet, "the capability cannot be taken away from us" is sometimes worth more than "the capability is slightly stronger this quarter," and it is a guarantee the closed model structurally cannot offer.

What Mistral Actually Does

Set the philosophy aside and look at the capabilities. In practice, Mistral gives you:

A frontier assistant in Vibe. Across its three modes, Vibe holds a conversation, writes and analyzes, researches the web, generates and runs code, and — in Work mode — acts as an agent that delegates multi-step tasks across your connected apps (Parts III and IV).

A family of open-weight models. From smaller, efficient models that run on modest hardware to large flagship models competitive with the best, all released as weights you can actually obtain and use (Chapter 4, Appendix A). As of this writing the flagship-class members are **Mistral Large 3** (the most capable, under the permissive Apache 2.0 license) and **Mistral Medium 3.5** (a strong, efficient model under a modified-MIT license, and the engine behind much of Vibe) — but model names and the exact lineup move fast, so confirm the current roster against Mistral's documentation.

A developer platform in La Plateforme. An API and tooling to build on Mistral's models, including fine-tuning to adapt a model to your own data and domain (Part VI).

The option to run it yourself. Download the weights and deploy them on your own infrastructure, from a workstation to a private cloud, with Mistral offering enterprise support for exactly that (Part V).

What ties these together is choice. With most AI providers you get one deployment model: their cloud, their terms. With Mistral you get a spectrum — hosted assistant, hosted API, private cloud, fully self-hosted — and you pick the point that fits your capability needs and your control requirements. That spectrum is the product.

It is worth walking the spectrum once from end to end, because the four points are not just "the same thing at different prices" — they trade control against convenience in ways that matter for the decision. At the most convenient end sits the **hosted Vibe assistant**: you log in, you type, Mistral runs everything, and you give up the most control in exchange for giving up the most effort. One step in is the **hosted API**, La Plateforme, where you still send your data to Mistral's servers but you are building your own application around the model rather than using the chat box — more control over how the capability is used, the same trust placed in where it runs. Further along is **private deployment**, where Mistral's models run on infrastructure dedicated to you, often inside your own cloud account or a managed enclave, so your data stays within a boundary you define while Mistral helps you operate it. And at the far end is **fully self-hosted**: you download the open weights and run them on your own hardware, answering to no one, sending nothing anywhere, owning the entire stack and the entire operational burden that comes with it.

Notice that the burden moves opposite the control. The hosted assistant asks almost nothing of you and gives you the least say; full self-hosting gives you total say and asks you to run, secure, update, and scale a frontier model yourself. Most users are happiest somewhere in the middle, and the right point is not a matter of taste but of what your data rules require and what operational capacity you actually have. The mistake to avoid is assuming you must pick the sovereign extreme to get the sovereign benefit — for many regulated users, Mistral's European hosted infrastructure clears the bar without their ever touching a GPU.

The point of the spectrum is that you match the deployment to the requirement, not that you always reach for the most hardcore option available.

Who's Using Mistral

Roughly three overlapping groups, in my experience.

Individuals and teams who want a capable, privacy-respecting assistant and like that it comes from a company aligned with European data values — using Vibe for the same everyday and professional work any assistant handles, with a lighter mind about where their data goes.

Developers and builders who want open-weight models they can run, fine-tune, and deploy on their own terms — for cost control, for data control, for the freedom to customize, or simply to avoid lock-in to a single closed provider (Part VI).

Regulated and sovereignty-conscious organizations — European institutions, regulated industries, the public sector, and any business for which data residency and self-hosting are requirements rather than preferences. For this group Mistral is often not one option among many but the *only* option that clears the bar, and Part V and Part VII are written with them squarely in mind.

This book serves all three, and the parts are arranged so you can read the one that is yours and navigate around the rest.

The Through-Line

If there is a single idea to carry out of this chapter, it is the one the hospital story opened with: the American flagships ask you to choose between capability and control, and Mistral was built to refuse the choice. Everything distinctive about it — the open weights, the permissive licenses, the self-hosting, the European data residency, the sovereignty framing — flows from that one refusal. The Vibe assistant is the friendly face of it; the open models underneath are the substance.

The rest of Part I fills in the picture — where Mistral came from and why Europe wanted its own champion, how Le Chat became Vibe, and what the model family actually contains. For now, the one-sentence version is enough to build on: Mistral is the capable AI you are allowed to control, and using it well starts with knowing which kind of control you actually need.

Chapter 2: The Mistral Story: Europe's AI Champion

In the spring of 2023, three researchers walked away from some of the best jobs in artificial intelligence to start a company in Paris that did not yet have a product, a model, or much of a plan beyond a thesis. The thesis was that the most important AI in the world was being built by a handful of American labs and a handful of Chinese ones, that Europe had the talent to build at the frontier but was on track to build none of it, and that this was a problem worth quitting your job over. A few months later the company they founded raised one of the largest seed rounds in European history on little more than that argument and the reputations of the people making it. The company was Mistral, and the argument it was founded on is the same argument that, three years later, explains nearly everything distinctive about the tool in your hands.

This chapter tells where Mistral came from, because the gap between Mistral and the American flagships is, more than anything, a gap in *why the company exists at all*. The US labs were built to reach the frontier. Mistral was built to reach the frontier *and keep it open and European*, and that second clause is not a marketing flourish bolted on afterward. It is the founding premise, and it shapes the product all the way down.

Three Founders and a Missing Champion

Mistral was founded in early 2023 by three French researchers who had been doing frontier-level work inside the American giants. The shorthand the press always reaches for is "ex-DeepMind and ex-Meta," and it is accurate: the founding team came out of Google DeepMind and Meta's AI research organization, which is to say out of exactly the labs whose dominance they set out to counter. These were not outsiders with a grievance. They were insiders who had helped build the thing they were now choosing to compete with, and who had concluded that the field's center of gravity was drifting somewhere they did not want it to settle.

The missing piece they saw was a European champion. Europe had produced an enormous share of the field's foundational researchers — many of the people staffing the American labs were European by training — but it had no frontier lab of its own. The talent was being exported; the models, the infrastructure, and the resulting economic and strategic weight were being built elsewhere. To a French researcher in 2023, the situation looked a little like watching your country's best engineers design someone else's aircraft. Mistral's founding bet was that this was both a problem and an opportunity: a problem for European independence, and an opportunity for whoever moved first to fill the gap.

I want to be careful not to tell this as a fairy tale, because the founding also rode a genuine wave of money and hype. The seed round was huge, the valuations that followed were enormous, and a fair amount of the early excitement was European institutions and investors wanting the champion to exist as much as it wanting to be a champion. Some of that enthusiasm was strategic and earned; some of it was the same froth that floated every AI company in those years. The honest version is that Mistral started with real talent, a real thesis, and a real tailwind of European appetite for an answer of its own — and that the appetite was part of the fuel from day one.

It is worth pausing on *why* a seed round that size was even possible on a thesis and a few resumes, because it tells you something about the moment. In early 2023 the field had just watched a single closed assistant cross from research curiosity into mass phenomenon, and every serious investor and every serious government suddenly understood that frontier AI was going to be strategically decisive and that almost none of it was being built in Europe. Into that realization walked three people who had demonstrably helped build the frontier elsewhere and were proposing to build it at home. The money was not betting on a product, because there wasn't one. It was betting on the combination of proven builders and a gap everyone had just noticed was dangerous. That is an unusual thing to be able to raise on, and Mistral could raise on it precisely because the timing put the gap in sharp relief. The lesson buried in the funding story is that Mistral's whole reason for being — *Europe has the talent but not the lab* — was not a pitch the founders had to sell so much as a fact the market had just been forced to confront.

The Sovereignty Thesis

The word that organizes the whole company is *sovereignty*, and it is worth slowing down on, because it means something more specific than national pride.

Sovereignty, in the sense Mistral cares about, is the ability to use powerful AI without surrendering control over your data, your infrastructure, or your dependence on a foreign provider. For an individual that can sound abstract. For a European hospital, bank, ministry, or manufacturer it is concrete and often legally binding. Under European data-protection rules, and under the instincts of regulators who watched the consumer internet get built and owned almost entirely outside Europe, the prospect of every important institution on the continent routing its sensitive data through a handful of American servers was not a convenience. It was a strategic exposure — to foreign law, foreign outages, foreign policy shifts, and foreign business decisions made with no European seat at the table.

Mistral's thesis was that AI did not have to be built that way. If a European company could field models capable enough to matter, released openly enough to be run on European soil under European rules, then institutions that could not or would not depend on the American clouds would finally have an option that cleared their bar. The point was never that European AI would be a little faster or a little cheaper. The point was that it would be *theirs* — runnable inside their own walls, governed by their own laws, free of the structural dependence that the closed American model builds in by default. Sovereignty is the answer to a question the US flagships do not even ask, because for most of their customers it never comes up.

It is worth separating the strands inside "sovereignty," because the word gets used loosely and the strands do not all apply to everyone. There is **data sovereignty** — the question of whether your information physically leaves your control and crosses into a jurisdiction whose laws you did not choose. There is **operational sovereignty** — the question of whether your ability to keep running depends on a foreign provider's uptime, pricing, and continued willingness to serve you. And there is **strategic sovereignty** — the largest, slowest strand, the question of whether a whole society's critical capability sits in the hands of powers that could, in a bad moment, withdraw it. A small European business mostly cares about the first. A bank cares about the first two. A government cares about all three at once, and cares about the third in a way that is less about any single contract than about not wanting a foreign capital to hold a quiet lever over the machinery of public administration. When Mistral says "sovereignty," it is making a single offer that happens to answer all three strands at different scales, which is part of why the word does so much work in the company's story — it is the rare pitch that lands for the sole trader and the ministry alike, for different reasons.

Open Weights as Strategy, Not Charity

Here is where the story turns from values into shrewd competitive logic, because Mistral's most distinctive choice — publishing its model weights — was not idealism. It was strategy, and a sharp one.

A smaller, younger, less-capitalized lab cannot win a head-on benchmark war with companies that have more compute, more researchers, and a multi-year head start. Mistral knew this. So instead of competing only on raw capability, where it would usually lose, it competed on a dimension the American giants had structurally chosen to abandon: openness. By releasing the actual trained weights of its models under permissive licenses — letting anyone download, inspect, run, fine-tune, and deploy them — Mistral made its models do things the closed flagships physically cannot. A closed model you rent over an API cannot be run inside a hospital's own network. A Mistral open-weight model can. That is not a benchmark advantage; it is a *capability* advantage that no amount of the competitors' compute can erase, because it is a property of how the model is licensed and shipped, not how smart it is.

The strategy did two things at once. It handed Mistral the sovereignty story on a platter — open weights are *how* you keep data on European soil — and it built a developer ecosystem fast, because researchers and builders the world over will reach for a capable model they can actually have over a slightly stronger one they can only borrow. Every developer who built on a Mistral open model, wrote a tutorial, fine-tuned a variant, or shipped a product on top of one was distribution the company did not have to pay for. Openness was the growth engine and the differentiator in a single move.

The mechanics of that growth engine are worth seeing clearly, because "openness builds an ecosystem" can sound like a slogan until you trace how it actually compounds. When a researcher can download the weights, they can publish a paper that uses the model, and that paper cites Mistral and teaches the next researcher how to use it. When a startup can run the model on its own servers, it can build a product without a per-call bill that scales with its success, and that product becomes a reference customer and a case study. When a hobbyist can fine-tune a variant on a weekend, they post it to a model-sharing hub, and now there are a dozen specialized Mistral derivatives that the company never had to build and that make the base model more useful to everyone who finds them. None of that happens around a closed model, because you cannot build a durable open ecosystem on a capability that can be withdrawn or repriced underneath you. Each of these moves costs Mistral nothing and leaves behind documentation, tooling, derivatives, and goodwill that a closed competitor would have to manufacture and pay for. That is the asymmetry the open bet exploits: a smaller company cannot outspend the giants on compute, but it can let thousands of outsiders do its evangelism, integration, and long-tail customization for free, simply by not standing in their way.

There is a second, harder-nosed edge to the strategy that the idealistic framing tends to skip. Openness is also a way to *commoditize the layer your competitors are trying to charge for*. If the American giants' business depends on renting access to a capable model, then a comparably capable model given away for free puts downward pressure on what anyone can charge for the model itself, and pushes the value — and the margin — up the stack toward the things Mistral can still sell: support, hosting, private deployment, enterprise features, and the assistant. A challenger that cannot win the model race can still change the rules of it, devaluing the prize its larger rivals are fighting over. That is a more aggressive reading of the open bet than "we believe in openness," and both readings are true at once. Mistral genuinely holds the values *and* the values happen to be the sharpest competitive weapon available to a company in its position. The two are not in tension; they are the same move seen from two angles.

I will not pretend the openness has been pure or unchanging, and Chapter 14 is the place where I lay out exactly what "open weight" does and does not mean, because the phrase is slipperier than it sounds and Mistral's own practice has shifted as its models grew more valuable. For the origin story, the durable point is this: Mistral chose openness deliberately, as the way a smaller European challenger could carve out ground the giants had ceded — and it worked.

The Climb to the Frontier

What followed the founding was a fast climb, and it is easy to lose in a blur of model names, so here is the shape rather than the changelog.

The early models punched above their weight. Mistral's first releases were smaller, efficient open models that startled people by competing with much larger ones. The reputation the company earned early was for doing more with less — squeezing frontier-adjacent capability out of models small enough to run on modest hardware. That efficiency reputation is still part of Mistral's identity and still shows up in the model family today (Chapter 4).

The models scaled up toward the frontier. Across successive generations Mistral moved from "impressively efficient" to "genuinely frontier-competitive," fielding larger flagship models that trade real blows with the best from the American labs on reasoning, coding, and knowledge. By the Mistral 3 family — the current generation as I write — dismissing Mistral as a lightweight is simply out of date.

The assistant grew up. Alongside the models, Mistral built a consumer-facing assistant, first called Le Chat and later renamed Vibe, that turned the underlying capability into something an ordinary person could use without touching code. That product arc is its own story, and Chapter 3 tells it.

I want to keep this honest, the way the rest of the book stays honest: Mistral is the smaller company in this fight. It has less compute than the American giants, a smaller ecosystem, fewer people, and it is not always — not even usually — the model topping the leaderboard on the day you check. Its climb has been real and fast, but it is a climb *toward* the frontier from a position of fewer resources, not a march at the front of the pack. Chapter 24 makes the comparison to the US flagships squarely and without flinching. The reason Mistral matters anyway is the reason this whole chapter has been building toward: it offers a combination — capable *and* open *and* sovereign — that the bigger players do not, and for a great many users that combination outweighs a few benchmark points.

Why Europe Wanted Its Own

Step back from the company and look at the continent, because Mistral did not rise in a vacuum. It rose because Europe had decided, more or less collectively, that it could not afford to sit out the most consequential technology of the era as a pure consumer of other people's models.

The reasons stack up and reinforce each other. There is **regulation**: Europe writes the world's most demanding rules on data and, increasingly, on AI itself, and a homegrown champion that builds with those rules in mind is worth more to European institutions than a foreign provider retrofitting compliance after the fact. There is **data**: the instinct, hard-earned from two decades of watching the consumer internet get owned elsewhere, that the continent's sensitive information should not all flow through servers it does not control. And there is **independence** in the plainest geopolitical sense — the unwillingness to be wholly dependent on American and Chinese AI for a capability that is starting to underwrite economic competitiveness, public administration, and national security alike. A Europe that rents all of its AI from two other powers is a Europe that has handed those powers a quiet veto over its own future. Mistral exists, and is celebrated as a champion, because enough of Europe decided that was a bargain not worth making.

That is the founding, told straight: three insiders who left to build the champion their continent lacked, a sovereignty thesis that answers a question the giants never ask, an open-weight strategy that turned a smaller company's disadvantage into a moat, and a fast but honestly uneven climb toward a frontier still mostly held by larger rivals. Hold those four threads, because the rest of the book is them playing out in practice — starting with the friendliest face of the company, the assistant that began life as Le Chat and grew into something the original name no longer fit.

Chapter 3: From Le Chat to Vibe

In its early days, Le Chat was the kind of assistant you treated exactly the way you treated every other one: you typed a question, it typed an answer, you copied the part you wanted, and you went back to doing the actual work yourself. The assistant was the smart part; the doing was still yours. Ask it to draft a paragraph, then paste the paragraph into the document. Ask it to summarize a report, then go act on the summary. It was a clever conversational partner that lived in a box and waited for you to ferry things in and out. That was what an AI assistant was, and Le Chat was a good one.

The thing that is now called Vibe does not work that way, and the name change is the company telling you so out loud. This chapter is about that arc — how Le Chat the chatbot became Vibe the work agent, why the rename in 2026 signals a real shift rather than a marketing refresh, and what it means that the tool is no longer waiting in its box for you to do the doing. It is the bridge between knowing what Mistral is and learning to actually use it, and it sets up Chapter 6, where the three modes get the full treatment.

Le Chat, the Chatbot

Le Chat launched as Mistral's consumer-facing assistant, and the name tells you the era it was born in. *Le chat* is French for "the cat," and a pun on "chat" in the chatbot sense — a light, slightly playful name for a light, conversational product. It did what assistants of its generation did, and did it well: it held a conversation, answered questions, drafted and edited text, explained things, reasoned through problems, and increasingly could look at an image or search the web when asked. If you used ChatGPT or Claude in those years, Le Chat was the familiar shape of that experience wearing Mistral's open models underneath and Mistral's European posture around the edges.

And it was a genuinely good version of that shape. Le Chat was fast — noticeably fast, because Mistral's efficiency reputation showed up as snappy responses — and it carried the company's distinctive advantages: capable models you could trust to come from a European company aligned with European data values, with the open-weight story humming underneath even if a casual chat user never thought about it. But for all that, Le Chat in its chatbot era lived inside the same fundamental limit every chatbot lived inside. It was an *advisor*, not an *actor*. It could tell you brilliantly what to do. It could not go do it. The boundary of the tool was the edge of the chat box, and everything past that edge — your documents, your inbox, your calendar, your other apps, the multi-step task that touched five of them — was on your side of the wall, your job to carry across.

The Shift From Answering to Doing

The whole industry crossed a line somewhere around 2025 and 2026, and Mistral's rename is its way of marking the crossing. The line is the difference between an assistant that *answers* and an assistant that *acts*.

An answering assistant is the model I described above: you bring it a question or a task, it produces text, you take the text and do the work. An *agentic* assistant — an agent — is one you can hand a goal rather than a question, and that will then plan the steps, reach into your connected applications, take the actions, and come back with the thing done rather than with instructions for doing it. The difference is not cosmetic. "Summarize these three reports, pull the figures that disagree, draft a one-page brief, and put it in the shared folder" is not a question with an answer; it is a *job* with several steps across several tools, and an answering chatbot can only ever hand you the pieces. An agent can take the whole thing.

Watch what that one request actually requires, because the gap between answering and acting is most visible when you slow the example down. An answering assistant, handed those three reports, can read them and produce a summary — but only if you have already pasted all three into the chat, and the moment it is done it hands the text back to you and forgets the rest of the sentence existed. It cannot open the shared folder, because it has no hands. It cannot reconcile the figures across the documents unless you have lined them up for it, because it cannot go fetch them. The instruction "and put it in the shared folder" is, to a chatbot, just more text to acknowledge politely and ignore. An agent treats the same sentence as a plan: locate the three reports wherever they live, read each one, extract and compare the figures, notice which disagree, compose the brief, find the shared folder, and write the file there — then come back and tell you it is done and where it put it. The chatbot gives you a paragraph and a to-do list you still have to execute. The agent gives you a finished file in the right place. Same words; entirely different relationship to the work.

The catch — and it is the catch that runs through the whole agentic turn — is that an agent earns its usefulness by being granted reach into your real applications, and reach is exactly what makes it consequential when it gets something wrong. A chatbot that misreads a report wastes a paragraph. An agent that misreads a report and then writes the wrong brief into the shared folder has touched the world, and you may not notice until a colleague does. This is not a reason to avoid the agentic mode; it is the reason the rest of the book pays real attention to what you connect, what you let an agent do unattended, and where you keep a human in the loop. The leap from answering to acting is genuinely valuable and genuinely worth understanding before you lean on it, which is most of why the modes get a chapter of their own.

This is the shift the rename announces. Le Chat was named for chatting, and as long as the product's center of gravity was conversation, the name fit. But once Mistral rebuilt it around delegating multi-step work — connecting to your apps, planning sequences of actions, executing them, and reporting back — "the cat," the chatbot pun, no longer described what the thing was. A name that says *chat* undersells a tool whose point is increasingly *work done on your behalf.* The company changed the name because the product had changed underneath it, and keeping the old name would have kept users thinking of it as the box-bound advisor it had outgrown.

Why "Vibe," and What the Name Signals

Vibe is a different kind of name, and the difference is the message. Where Le Chat named the *mechanism* — chatting — Vibe names something closer to the *relationship*: the sense of a capable collaborator you hand things to and trust to get the gist, the intent, the vibe of what you want and run with it. It is a name for a tool you delegate to rather than a tool you interrogate. Whether you love the name or find it a touch try-hard, its function is clear enough: it is meant to signal that this is no longer a question-and-answer box but a working partner.

I will be honest that names like this carry a whiff of marketing, and I am not going to oversell a rebrand as a revolution. The underlying models did not transform on the day the name changed; the capability built up over many releases, and the rename is a marker planted along the way, not the moment of transformation. But the marker points at something real. The honest read is that "Le Chat to Vibe" is the public, consumer-facing label on a shift that had been building underneath for a while — from a smart conversationalist to an agent that does multi-step work — and the new name is Mistral asking you to update your mental model of what the tool is for.

The Three Modes, Introduced

The clearest expression of the new Vibe is that it is organized around three modes, each a different posture for a different kind of work. I will introduce them here and leave the full treatment to Chapter 6, because choosing the right mode is most of using Vibe well and it deserves its own chapter.

Work mode is the agentic heart of the rename — the mode where Vibe stops answering and starts doing. In Work mode you hand Vibe a multi-step task that may span several of your connected applications, and it plans the steps, takes the actions, and returns with the job advanced or finished. This is the mode that "the cat" no longer described and "Vibe" is meant to capture.

Code mode is the developer surface: Vibe oriented around writing, reading, and changing software, reachable from a command line, inside a code editor, or on the web. It is built for the particular shape of programming work, where the agent reads a codebase, makes changes, and runs things. Part IV of this book is largely about it.

Chat mode is the classic experience, kept and not discarded — the turn-based conversation that Le Chat always was, for when what you want genuinely is a question answered, a thing explained, a paragraph drafted. The arrival of agents did not make conversation obsolete; plenty of the time a straight answer is exactly the right tool, and Chat mode is there for it.

The thing to notice is that the three modes are not three products bolted together. They are three postures of one assistant, and the skill the rest of this book builds is knowing which posture a given piece of work wants.

A quick way to internalize the difference before Chapter 6 takes it apart properly: the modes differ less in *what* they can do than in *how much initiative they take and how much reach they have*. Chat mode is the lowest-reach posture — it works with what you give it and hands the result back, touching nothing else. Work mode is the highest-reach posture — it is allowed to plan, to span your connected apps, and to act, which is why it is the one you delegate to and also the one you supervise most carefully. Code mode is reach pointed at a specific domain, software, where the actions are reading and changing and running code rather than touching your inbox and calendar. If you find yourself unsure which to use, the honest rule of thumb is to ask how much you want the assistant to *do on its own*: if the answer is "just tell me," that is Chat; if the answer is "go handle the whole multi-step thing," that is Work; if the work is a codebase, that is Code. You will refine this with practice, and the wrong choice is rarely catastrophic — using Chat for something Work could have automated just means you did a few steps by hand, which is exactly how the old chatbot always worked.

What Existing Users Kept

If you used Le Chat and worried that the rename meant your familiar tool got taken away and replaced with something stranger, the reassuring news is that almost nothing was subtracted. The transition was additive. Everything Le Chat did, Vibe still does — the conversation, the writing, the analysis, the web search, the speed, the European posture, the open models underneath. Chat mode *is* Le Chat, preserved and carried forward under a new roof. Your old way of working did not break; it became one of three things the tool can now do.

What changed is that two larger capabilities grew up around that familiar core. The agentic Work mode arrived for the multi-step jobs the chatbot could only ever advise on, and the developer-focused Code mode matured for software work. So the right way to understand the rename, as a returning user, is not "they replaced my chatbot" but "they kept my chatbot and built an agent and a coding tool around it, and renamed the whole thing to reflect that the chatbot is now the smallest part of what it can do." You lost nothing and gained two new postures, and the only real adjustment is mental: learning to reach for Work mode when a job is bigger than a conversation, instead of falling back on the old habit and ferrying every step across the wall by hand.

The Underuse Trap

I want to dwell on that adjustment, because it is the single most common way returning Le Chat users get less out of Vibe than they could — and it is a quiet failure, the kind you never notice because nothing breaks. The tool works fine; you simply keep using a fraction of it. The trap is muscle memory. After months or years of treating an AI assistant as an advisor in a box, the habit of ferrying work across the wall by hand becomes invisible. You ask Vibe to draft the email, then you copy the draft into your mail client and send it yourself. You ask it to summarize the three reports, then you go reconcile the figures and write the brief and save the file by hand, exactly as you always did, never noticing that you have just performed by hand a job you could have delegated whole. The chatbot trained you to do the doing, and the training does not undo itself the day the name changes.

The tell that you are stuck in the trap is simple: notice how often you copy something out of the assistant and then carry out the next step yourself. Every one of those moments is a place where the old advisor model is showing through, and many of them are places where Work mode could have taken the whole sequence. The fix is not to delegate everything — plenty of work genuinely is a quick question that Chat answers best, and forcing every trivial ask through an agent would be its own kind of mistake. The fix is to develop a reflex to *ask* the question: is this a thing I want told to me, or a thing I want done? When the honest answer is "done," and you catch yourself reaching for copy-paste anyway, that is the trap, and naming it is most of escaping it.

The deeper point is that the rename is asking you to change a habit, not just learn a feature. Features you pick up by reading the manual. Habits you only change by noticing the old one in the act and deliberately choosing the new one until it sticks. That is why this chapter spends its energy on the *shift* rather than on a tour of buttons — the buttons are easy and Chapter 6 covers them, but the shift from "advisor I consult" to "partner I delegate to" is the thing that determines whether Vibe is a better chatbot or a genuinely different tool in your hands.

That is the arc — from a clever advisor in a box to a working partner you delegate to, marked by a name that traded *chat* for *vibe* because the product had outgrown the smaller word. Knowing the history is enough to stop you from underusing the tool, which is the most common mistake returning Le Chat users make: treating Vibe as the old chatbot and never once handing it a real job. The mechanics of all this — the model family that powers it, the plans that gate it, the three modes in working detail — are what the rest of the orientation and Part II lay out.

Chapter 4: The Mistral Model Family

Everything Vibe does — the conversation, the agentic work, the coding, the analysis — is the work of a model, and the model is where the capability ultimately lives. You can use Vibe well without thinking about the models much, the same way you can drive without knowing what is under the hood. But with Mistral, a little understanding pays off more than it does elsewhere, because the models are not just the engine — they are the *product you can take home*. The open weights that make Mistral Mistral are model facts, and the choices the company makes about which model gets which license are the sovereignty story made concrete. This chapter is that understanding, and it closes Part I. One caution before we start: this is the most perishable chapter in the orientation. Model names, license terms, context sizes, and prices change faster than anything else in this book. I will give you the durable shape and flag every specific.

A Family, Not a Single Model

The first thing to fix in your mind is that "Mistral" is not one model. It is a *family*, and the family framing is not a technicality — it is the whole design. Where some providers point you at a single flagship and call it their AI, Mistral fields a range, from smaller efficient models that run on modest hardware up to large flagship models competitive with the best, and the range exists precisely so that you can pick the point that matches your needs for capability, cost, and — uniquely here — how openly the model is licensed and where you are allowed to run it.

The current generation as I write is the **Mistral 3** family, and two members anchor it: a most-capable flagship and an efficient workhorse. I will take them in turn, but hold the family idea above the individual names, because the names are exactly the kind of detail that drifts. **Confirm the current roster against Mistral's own documentation** before you rely on a specific version — the live page is always the authority, and I would rather you check than trust a number that has aged on a printed page.

Mistral Large 3: The Flagship

The most capable member of the family as I write is **Mistral Large 3**, and it is the model to reach for when capability is what matters most. A few facts about it are worth carrying, with the standard caution that the specifics are perishable.

It is large, and built to be efficient about it. Mistral Large 3 is a sparse mixture-of-experts model — in the neighborhood of 675 billion total parameters, but "sparse," meaning that for any given request only a fraction of those parameters actually fire. That is how a model this large stays practical to run: it has the knowledge capacity of something enormous but the working cost of something much smaller, because it does not light up the whole network for every token. This is the architectural trick that lets Mistral field a frontier-scale flagship without the resources of the American giants, and it is worth knowing the term — mixture-of-experts, MoE for short — because you will see it across the field.

The mental picture that makes mixture-of-experts click is a large hospital with many specialists rather than a single enormous generalist. When a case comes in, you do not wake every doctor in the building; a triage step routes the case to the handful of specialists it actually needs, and the rest stay idle. An MoE model works the same way: the network is divided into many "expert" sub-networks, and for each chunk of text a small routing component decides which few experts are relevant and activates only those. The model holds the combined knowledge of all its experts — which is why the total parameter count is so high — but pays only for the few it consults on any given token, which is why it runs far cheaper than a dense model of the same nominal size would. The trade is real, not free: an MoE model needs enough memory to *hold* all those experts even though it only runs a few at a time, so it is hungry for the kind of high-memory hardware that makes self-hosting the very largest models a serious undertaking. But for the cost that matters most in practice — the compute spent per request — sparsity is what turns "frontier-scale" from a research luxury into something a company can actually afford to run, and it is the single most important reason a smaller European lab can field a flagship at all.

It is the most open of the family, under Apache 2.0. This is the part that matters most and the part no American flagship matches. Mistral Large 3's weights are published under the **Apache 2.0 license** — one of the most permissive open licenses there is, with essentially no strings on who can use it or how. You can download it, inspect it, run it on your own hardware, fine-tune it, and deploy it commercially, without asking anyone's permission and without sending anyone your data. That a company's *most capable* model is also its *most openly licensed* is a deliberate and pointed choice, and the clearest possible expression of the sovereignty thesis from Chapter 2: capability and control are not a trade-off here. Chapter 14 unpacks exactly what Apache 2.0 and "open weight" do and do not mean — don't take this paragraph as the full treatment — but the headline is that the flagship is genuinely, permissively open.

Its API pricing is competitive. For programmatic use through La Plateforme (Chapter 18), Mistral Large 3's token pricing as I write is in the neighborhood of **$0.50 per million input tokens and $1.50 per million output tokens** — input being what you send the model, output being what it generates back. That ratio, output costing more than input, is standard across the industry, and these rates put the flagship in the competitive-to-aggressive range for a frontier model. **Verify both figures against Mistral's current API pricing before you budget against them** — rates move, they vary by model within the family, and the printed number here is a reference point, not a quote.

Mistral Medium 3.5: The Workhorse

The flagship is not the model most people actually touch the most. That distinction belongs to **Mistral Medium 3.5**, the efficient workhorse of the family, and — importantly — the engine behind much of what Vibe does day to day. When you have a routine conversation in Vibe, odds are good you are talking to a model in this class rather than the flagship, because most work does not need the heaviest model and Medium is fast, cheap, and more than capable enough.

It is a smaller, denser model tuned for efficiency. Mistral Medium 3.5 is in the neighborhood of 128 billion parameters and "dense" rather than sparse — every parameter participates in every request — and it is described as a *merged* model, blending capabilities into a single efficient package. The practical upshot is a model that delivers strong capability at a fraction of the flagship's cost and latency, which is exactly what you want as the default engine behind an assistant that answers all day.

The dense-versus-sparse contrast is worth holding onto, because it explains why both models exist rather than one obviously dominating. A dense model runs every parameter on every token, so a 128-billion-parameter dense model does roughly 128 billion parameters' worth of work each time — predictable, uniform, and at this size genuinely fast, with no routing overhead and no need to hold a vast bench of idle experts in memory. The sparse flagship holds far more total knowledge but only consults a slice of it per token. So the honest way to think about the two is not "bigger is better" but "different shapes for different jobs." The flagship wins when a task genuinely needs its deeper reservoir of capability — hard reasoning, gnarly code, the long tail of obscure knowledge. The workhorse wins when a task is well within reach of a strong mid-sized model and what you actually care about is answering quickly and cheaply, which describes the overwhelming majority of everyday assistant traffic. That is precisely why Medium sits behind so much of Vibe: most of what people ask an assistant all day does not need the heaviest model, and putting the flagship on every trivial question would be slower and more expensive for no benefit the user would notice. The family ships both because the right answer to "which model" is "it depends on the request," and a good system routes accordingly.

Its license is open but more conditional. Here is a nuance that matters and that Chapter 14 will treat in full: Medium 3.5 is released under a **modified-MIT license** rather than the unconditional Apache 2.0 of the flagship. MIT is itself a famously permissive open license; the *modification* is the catch. The modified terms carve out exceptions for very-large-revenue firms — roughly, the biggest companies have to come to a commercial arrangement rather than using it freely. For the overwhelming majority of users this changes nothing; you can use it as freely as any open model. But it is a real and meaningful difference from the flagship's no-strings Apache 2.0, and it is a small, concrete example of the larger truth this book keeps returning to: Mistral is *meaningfully more open than the American flagships, with real and shifting limits that vary by model.* Don't carry away "everything Mistral makes is free for everyone." Carry away "the license depends on the model, and you should check."

The Smaller Models

Below the flagship and the workhorse, Mistral fields smaller, efficient open models — the kind that made the company's early reputation for doing more with less, small enough to run on genuinely modest hardware. I am not going to enumerate them by name, because that is the fastest-drifting part of the whole lineup, but the durable point is that the family extends downward into models you can run on a single workstation or a tight budget, where the trade is some capability for a lot of efficiency and the freedom to run the thing almost anywhere. For developers and self-hosters especially, these smaller members are often the right starting point, and Part V and Part VI return to them.

It is easy to dismiss the small models as the consolation tier — the ones you settle for when you cannot afford better — and that instinct is usually wrong. For a great many real jobs, a small model is not a compromise but the *correct* choice, because the job does not need frontier reasoning and does need to be fast, cheap, private, and runnable somewhere a big model cannot go. Classifying support tickets, extracting fields from a stream of invoices, drafting routine first-pass replies, tagging documents, powering a feature inside an app that has to respond in milliseconds — these are high-volume, well-bounded tasks where throwing the flagship at every request would be slower, far more expensive, and no more correct. A small model that you can run on a single machine inside your own network, with no per-call bill and no data leaving the building, often beats a brilliant hosted model on every axis that matters for that particular use. The skill is matching the model to the task honestly, and Chapter 19 is built entirely around that judgment. For now, carry the right mental frame: the small models are not the bottom of a quality ladder but a different point on a capability-versus-cost-versus-deployability curve, and for the right job they are exactly where you want to be.

A Worked Example: Choosing Within the Family

To make the family idea practical rather than theoretical, walk through how a single team might use three members at once, because in real deployments you rarely pick one model and stop. Picture a small legal-tech company building a tool that helps firms review contracts.

For the **high-volume, low-stakes** layer — scanning every incoming document and tagging it by contract type, flagging which ones mention specific clauses — they reach for a **small model**, run on their own servers. It is cheap enough to throw at thousands of documents a day, fast enough to keep the queue moving, and it keeps client contracts inside the company's own network, which their customers' compliance teams require. The work is bounded and repetitive; a frontier model here would be expensive overkill.

For the **everyday interactive** layer — the lawyer asking the tool to summarize a clause in plain language, or to compare two versions of an agreement — they use a **Medium-class workhorse**. It is responsive enough for a back-and-forth session, capable enough to handle the nuance of legal prose, and economical enough to sit behind every user all day. This is the same calculus that puts Medium behind much of Vibe.

And for the **hard, high-stakes** layer — the request to reason through how a set of interlocking clauses would interact under a specific scenario, where a wrong answer is genuinely costly — they escalate to the **flagship**, accepting its higher latency and cost because the difficulty of the task justifies them. They do not route everything to the flagship; they route the small fraction of requests that actually need it.

That is the family doing what it is designed to do: not one model for all work, but a graduated set you assign by the shape of each job. The team also gets a sovereignty bonus that no closed provider could match — the high-volume layer that touches the most client data runs entirely on infrastructure they control, while only the harder, rarer requests reach out to a hosted flagship if they choose. The family structure is what makes that split possible at all.

Context: How Much a Mistral Model Holds at Once

A model's *context window* is how much text it can keep in mind in a single conversation — your prompt, the documents you have given it, and everything said so far, all counted together. When you exceed it, the earliest material falls out of the model's view, which is why a very long session can seem to "forget" how it began.

The flagship-class members of the Mistral 3 family offer a **large context window — on the order of 256,000 tokens** — which is a great deal of room. Long documents, large bodies of reference material, and extended sessions all fit comfortably where they would have overflowed an earlier model. Roughly speaking, a few hundred thousand tokens is several long books' worth of text held in mind at once, so for ordinary work you rarely have to think about the limit: you can paste the whole report, share the long thread, hand over the big file, and the model keeps it in view. Treat the exact figure as a check-the-docs number — context sizes change and sometimes differ by model and by surface — but the durable point holds: this generation has room to spare for long-context work.

Why the Licensing *Is* a Feature

It is tempting to read the last several pages as housekeeping — model names, parameter counts, license abbreviations — and skim. Don't, because the licensing is not housekeeping here. It is the feature.

For every closed American flagship, the license question does not arise, because there is no license; there is only an API and terms of service, and the model never leaves the provider's servers. Mistral's decision to publish weights under named open licenses is what converts "a capable model" into "a capable model you are allowed to run yourself, on your own soil, under your own rules." That is the entire sovereignty proposition from Chapter 1 and Chapter 2, and it lives in exactly these unglamorous license strings. Apache 2.0 on the flagship is *why* a European hospital can run Mistral's best model inside its own walls. The modified-MIT on Medium is a smaller, conditioned version of the same gift. The licenses are not fine print under the product. With Mistral, they substantially *are* the product, and the rest of this book — especially Part V — is what you do with that.

What to Carry Forward

You do not need to memorize a parameter count or a license clause to use Vibe well. What is worth carrying out of this chapter is the shape of the thing.

A **family, not a single model** — the Mistral 3 generation, ranging from small efficient models up to a frontier flagship — so you can pick the point that fits your capability, cost, and openness needs.

A **most-capable flagship, Mistral Large 3** — a large sparse mixture-of-experts model under the unconditionally permissive Apache 2.0 license, the clearest proof that capability and openness need not trade off.

An **efficient workhorse, Mistral Medium 3.5** — a smaller dense model under a modified-MIT license with carve-outs for the largest firms, fast and cheap and the engine behind much of Vibe.

A **large context window**, around a quarter-million tokens at the top end, big enough that long documents and long sessions just fit.

And a set of **numbers that are real but perishable** — names, parameter counts, license terms, context sizes, and the API rates above — that you confirm against Mistral's live documentation rather than trusting on a printed page.

That is the engine, and with it Part I has done its job. You now know what Mistral is, where it came from, how Le Chat became Vibe, and what powers the whole thing. What you do not yet know is how to get at it — the plans, the tiers, the surfaces Vibe lives on — and that is where Part II begins.

Part II: Getting Started with Vibe

Audience: everyone. The setup is easy — but read Chapter 5 before you spend anything, and note that the Vibe subscription and the API are separate purchases.

"Most of using Vibe well is choosing the right mode for the job in front of you. The rest follows from that one decision."

— A working principle

Chapter 5: Access and Plans

Spend ten minutes on Vibe's plans before you build anything on top of them, and you will save yourself the mistake I am about to talk you out of. The first time I put Vibe in front of a small team, I let one person sign up for the paid plan, get excited, and build a whole workflow around a connector that pulled from their spreadsheets — and then discovered, three weeks later, that the agentic job they wanted to schedule was sized for a shared workspace none of us had. We did not need a more capable model. We needed to have understood the plan ladder before we started building habits on top of it. So that is what this chapter is: ten minutes, spent up front, on where Vibe lives, what the plans actually unlock, and the one billing fact that surprises almost everyone. After this, the rest of Part II is just using the thing.

A warning that applies here with unusual force. **Plan names, prices, and the exact boundary lines between tiers are the most perishable facts in this book.** Mistral is a fast-moving European company in a fast-moving market, the assistant was renamed from Le Chat to Vibe inside the last year, and the plan structure has already shifted more than once. I am going to give you the *shape* of access, which is durable, and I will give you the mid-2026 euro prices because concrete numbers are more useful than coy vagueness — but every one of them carries a flag: verify it against Mistral's own pricing page, in your own currency, on the day you decide to pay. Treat the figures below as a map drawn at a particular moment, not as the territory.

Where You Actually Use Vibe

Before the money, the surfaces. Vibe is not a single app; it is a small constellation, and which surface you reach for shapes what it can do.

The web. The primary home is the browser. Go to Mistral's Vibe site, sign in, and you have the full assistant — all three modes (Work, Code, and Chat, the subject of Chapter 6), your Libraries, your Connectors, your Memory. If you only ever use one surface, this is the one. It is also where the agentic Work mode is most complete, because acting across your connected business apps is a job better suited to a real screen than a phone.

Mobile. There are iOS and Android apps, and they are not toys — Chat and Work both live there, so you can hand Vibe a multi-step task from your phone and let it run. What mobile is genuinely good at is starting and checking on work: kick off a job in the morning, glance at the result over lunch. Deep, screen-heavy work still belongs on the web.

The developer surfaces. Vibe's Code mode reaches you as a CLI tool you run in your terminal, a VS Code extension inside your editor, and on the web. As of roughly June 2026 you can even start a coding job from Slack. These are the developer's doors into Vibe, and Part IV is built around them; if you write software, Chapter 13 is where they get their due.

The thing to hold onto is that **one account, one subscription, follows you across all of these.** You are not buying the web app and the mobile app separately. You are buying access to Vibe, and you reach it from whichever surface fits the moment.

That single-account fact has a practical consequence worth naming, because it shapes how a real day with Vibe actually flows. You can begin a piece of work on one surface and finish it on another. Picture a Monday: on the train you open the mobile app and kick off a Work job — "pull last week's support tickets from the connected tool and draft me a summary." It runs while you finish your commute. You walk to your desk, open the web app, and the same job, the same conversation, the same Library and Connector context are all there waiting; you review the draft on a full screen, refine it, and send. Nothing was re-explained, nothing was re-uploaded, no second login. The work followed you. If you have ever lost a thread because you started something on your phone and could not pick it up on your laptop, the one-account model is the quiet thing that removes that friction, and it is worth structuring your habits around: start where you are, finish where the screen suits the task.

The Plan Ladder

Here is the ladder as of mid-2026, from least to most, with prices flagged for verification.

Free. There is a genuinely useful free tier. You get the latest models, basic web search, and basic Code capability, with a daily message ceiling that lands somewhere around twenty-five messages a day (verify — this limit moves). It is enough to feel how Vibe responds, run real questions, try a mode, and decide whether the tool earns a place in your week. It is the right place to *start* and the wrong place to *settle* if Vibe becomes something you lean on, because the daily ceiling will start to bite the moment you use it seriously.

Pro — about €14.99 per month (verify). This is the tier most individuals who rely on Vibe will want. It raises the usage limits substantially, opens up deeper reasoning, unlocks deep research over the web, and gives you meaningfully more capacity for Vibe's Code work. There is also a verified-student rate around €5.99 per month (verify, and expect to prove your enrollment). If Free starts feeling like a leash, Pro is the obvious next rung, and for a single working professional it is usually the right long-term home.

Team — about €24.99 per user per month (verify). This is Pro's logic extended to a group. You get a shared workspace, administrative controls, and more storage — the things a handful of people working together actually need: shared Libraries, shared Connectors, a place where the work is collective rather than trapped in one person's account. My team's early stumble was exactly this: we wanted shared, agentic workflows and tried to build them on individual Pro seats. If more than one person will touch the same knowledge and the same connected apps, Team is where that becomes sane.

Enterprise — custom, through sales. At the top sits the tier where Mistral's whole reason to exist comes forward: custom terms, self-hosted and private deployment options, and the data-residency and sovereignty guarantees that Part V and Chapter 17 are built around. This is not a checkout-page purchase; it is a conversation with Mistral's sales team, priced to your organization. If your interest in Mistral is the *sovereign* story — running models inside your own walls, keeping data under your own law — this is the door, and the rest of this Part is only the warm-up to it.

The honest summary, names and numbers aside: **Free to try, Pro to rely on it solo, Team to share it with a few people, Enterprise to control where it runs.** That structure is stable even as every euro figure around it drifts.

What Actually Changes as You Climb

It is worth being concrete about what moving up a rung *buys* you, because "higher limits" undersells it and "more features" oversells it. The honest middle is that each step up changes a specific thing about how hard you can lean on Vibe, and knowing which lever each rung pulls is how you tell whether you actually need the next one.

Free to Pro is mostly about ceilings and depth. The free tier's daily message cap is a real wall, and the first thing Pro does is push it far enough back that you stop thinking about it. But the more interesting unlock is *depth*: deeper reasoning on hard questions, and deep research over the web — the capability where Vibe will go out, read many sources, and assemble a synthesized answer rather than a single quick reply (Chapter 10). If you have been on Free and the thing that frustrates you is hitting the daily cap, Pro fixes that mechanically. If the thing that frustrates you is that the answers feel shallow on genuinely hard questions, Pro fixes that too, and that second reason is the better one to upgrade for.

Pro to Team is not about more power, it is about *shared* power. This is the rung people misjudge most often, because the instinct is "Team is just Pro for more people, so we'll buy Pro seats and save money." That instinct is exactly the mistake I confessed to opening this chapter. The thing Team gives you is not a bigger version of Pro — it is a *shared workspace*: shared Libraries, so the whole group draws on the same institutional knowledge instead of each person re-explaining it; shared Connectors, so a workflow built once is available to the team rather than trapped in one account; and administrative controls, so someone can actually manage who has access to what. Three people on three separate Pro accounts cannot share a Library or a Connector. The moment the *work* is collective — same documents, same connected apps, same standing jobs — individual seats stop being a saving and start being a structural dead end.

Team to Enterprise is about where the thing runs. Everything below Enterprise is hosted by Mistral on Mistral's terms; you are choosing how much of the hosted assistant you get. Enterprise changes the question entirely, from "how much access do I get" to "where does the model live and under whose law does my data sit." That is a different conversation — self-hosting, private deployment, data-residency guarantees — and it is the one Part V is written for. If your reason for being interested in Mistral at all is the sovereignty story, the subscription ladder below Enterprise is essentially a familiarization exercise before the conversation that matters.

The reason to hold this map in your head is that it tells you *why* to climb, not just *that* you can. People hit a wall on Free and reflexively reach for the most expensive plan they can justify, when the right move is to name the specific thing that is hurting — the cap, the depth, the sharing, or the deployment — and buy exactly the rung that fixes it.

The One Billing Fact Everyone Misses

Here is the thing I most want you to carry out of this chapter, because it has tripped up nearly everyone I have watched start with Mistral.

A Vibe subscription does not include API credits. Vibe — the assistant, in all its plans — and **La Plateforme**, Mistral's developer API, are *separately billed.* Paying for Pro, or Team, or even Enterprise, buys you the assistant. It does not put a single credit in your API account. The API is its own product with its own meter: you pay per use, by the token, to call Mistral's models from your own software. The two live in different parts of Mistral's world and they bill independently.

This matters because the two are easy to conflate. You will read, later in this book, about building on Mistral's models directly — and you might reasonably assume your Pro subscription covers a little of that. It does not. The moment you move from *chatting with Vibe* to *calling the API from your own code*, you are on a different bill entirely. Part VI is where the API gets its full treatment, including its costs (Chapter 21). I flag it here only so you do not build a plan around a budget that quietly has two line items where you thought it had one.

Usage Is Metered Even Inside a Plan

A smaller surprise, related to the last one: paying for a plan does not buy unlimited use. Within every tier there are limits, and the more demanding the work, the faster it draws them down. A quick Chat question costs almost nothing. A deep research run across the web, a long agentic Work job that scans an inbox and pulls spreadsheet data and assembles a report, a heavy Code session in a cloud sandbox — these consume far more of your allowance, and you can reach a tier's ceiling faster than light experience would suggest.

This is awareness, not alarm. If you keep hitting limits, the real question is whether you are leaning on the expensive capabilities — deep research, agentic

Work, heavy Code — enough to justify the next rung, or whether a lighter touch keeps you comfortable where you are. Either way, the meter runs, and the agentic features run it fastest.

It helps to have a rough mental model of the cost gradient, because it is steeper than newcomers expect. A plain Chat question — "rewrite this paragraph," "what's the difference between these two terms" — sits at the cheap end; you could ask hundreds in a day without noticing. A deep research run is perhaps an order of magnitude heavier, because behind the single answer Vibe is reading and reasoning over many sources. An agentic Work job that scans an inbox, opens several spreadsheets, synthesizes the contents, and writes a report is heavier still, because it is effectively a long chain of those operations stacked end to end. And a scheduled Work job multiplies that by however often it runs — a daily briefing is the same heavy job firing seven times a week, every week, whether or not you read the output that morning.

That last point is the one that quietly catches people. The natural reflex once Work mode clicks is to schedule everything: a morning digest, a weekly report, an hourly watch on some inbox. Each one is reasonable alone. Together they are a standing draw on your allowance that runs in the background whether you are paying attention or not. The discipline is not to avoid scheduled jobs — they are one of the best things about Work mode — but to schedule the ones you genuinely act on and to retire the ones whose output you have started skimming past. A digest you no longer read is the purest form of a meter running for nothing.

Which Plan Should You Start On?

A practical recommendation, with the standing caveat that you should confirm current terms before acting on any of it.

If you are **curious and unsure**, start on **Free.** Run real questions, try Work and Chat, see whether Vibe earns its keep before you pay anything. The daily ceiling is exactly the right amount of friction for a trial.

If Vibe becomes **a tool you reach for most days**, move to **Pro.** The higher limits, deeper reasoning, and deep research are what turn it from a curiosity into a working instrument, and for one person that is the natural home. Students should check for the verified rate before paying full price.

If **two or more of you will share knowledge and connected apps**, go to **Team** rather than stacking individual Pro seats. The shared workspace and admin controls are the whole point, and trying to fake them with separate accounts is the mistake I already confessed to.

If your real interest is **running Mistral on your own terms** — self-hosting, data residency, sovereignty — your path runs through **Enterprise**, and the subscription ladder is mostly a way to get familiar before that conversation. Part V is written for you.

And if you are a **developer planning to build on the models** rather than chat with the assistant, remember the billing fact above: the plan ladder may be beside the point. Your meter is the API's, and Part VI is your chapter.

Get that settled — a surface you can reach Vibe on, a starting plan that fits, prices you confirmed rather than took on faith, and a clear head about the API being a separate bill — and the spending decisions are behind you. What is left is the good part: actually using the thing, beginning with the choice that shapes everything else about working in Vibe — which of its three modes you are in.

Chapter 6: The Three Modes: Work, Code, and Chat

I watched someone fight Vibe for twenty minutes once, and the whole fight came down to a mode. She wanted it to go through her shared inbox, find every message from a particular client, pull the numbers they had sent across three different spreadsheets, and assemble a one-page status summary and she was typing all of that, patiently, into Chat. Chat answered the way Chat answers: helpfully, conversationally, by explaining how she *could* do those things. What she wanted was for Vibe to actually *do* them. She did not need a better prompt. She needed Work mode. The instant she switched, the same request became a job Vibe went off and performed, and her frustration evaporated. That is the chapter in one anecdote: in Vibe, the mode you are in is not a setting, it is the single decision that most shapes whether the tool does what you meant.

Most assistants give you one box and one behavior. Vibe gives you three, and they are genuinely different tools sharing a face. Choosing among them well is, more than any prompting trick, what it means to use Vibe skillfully. So let me make the distinctions concrete.

Chat: The Conversation You Already Know

Chat is classic, turn-based conversation — the thing every assistant does, done well. You type, Vibe answers, you follow up, it refines. It holds the thread, so you can steer across many turns without re-explaining. It can search the web when a question needs current information. It writes, summarizes, explains, reasons, drafts, brainstorms, and talks you through a problem.

What defines Chat is that it stays *inside the conversation*. It produces words back to you — an answer, a draft, an explanation, a plan — and then it is your move. It does not reach out and touch your other applications, and it does not go off and run a multi-step job on its own. It is the sharp, fast colleague at the next desk who will think through anything with you but hands the result back for you to act on.

Reach for Chat when the work is *thinking and producing text*. "Explain the difference between two contract structures." "Draft a warm reply to this email."

"Summarize this report and tell me what's missing." "Help me brainstorm names for a product." "Talk me through whether this plan has a hole in it." These are conversations, not jobs, and Chat is exactly the right tool for them. It is also where most people should start, because it asks nothing new of you — it is the assistant experience you already have an instinct for.

There is a subtler reason Chat is the right default, and it is about safety rather than gentleness. Because Chat stays inside the conversation and hands every result back to you, *nothing happens without your next move*. It cannot send an email you did not approve, cannot overwrite a file, cannot trigger anything outside the chat box. That makes it the ideal place to do the thinking *before* an action — to draft the email, argue with the draft, and get it right — and then act on the result yourself. A useful pattern that experienced Vibe users settle into is to do the composition in Chat, where it is safe to be messy and iterative, and to switch to Work only once they know exactly what they want done. Chat is where you decide; Work is where you delegate. Keeping the deciding in the mode that cannot act on its own is not timidity, it is good sequencing.

One honest limit of Chat, so the boundary is clear from both sides: because it does not reach into your applications, anything it says about *your* specific data is only as good as what you have pasted in. Ask Chat "what's the status of the Acme account" with nothing attached and it cannot know — it will either tell you it cannot see your systems or, worse on a bad day, guess. The fix is either to paste the relevant material into the conversation or, when the answer genuinely requires reaching live systems, to recognize that you have left Chat's territory and crossed into Work's. That recognition is the whole skill this chapter is teaching, and it cuts both directions.

Work: The Agent That Acts For You

Work is the agentic mode — it delegates and performs multi-step tasks across your connected business apps. This is the mode that makes Vibe feel like a different category of tool, and it is the one the woman in my opening story actually needed. In Work mode, Vibe does not just tell you how to do something; it goes and does it, in sequence, across the applications you have connected to it.

What it can do, concretely: scan an inbox you have connected and find what matters in it; pull data out of spreadsheets; gather information from across several connected tools and synthesize it; build a report from the pieces; and — this is the part that turns it from a clever trick into infrastructure — schedule recurring jobs, so a task you would otherwise repeat by hand runs on its own on a cadence you set. It works on the web and on mobile, which means you can hand it a job from your phone and let it churn.

The thing that makes Work possible is the trio of features Chapter 8 is devoted to: **Connectors** link Vibe to your apps so it has somewhere to act; **Libraries** give it saved knowledge to draw on; and **Memory** lets it carry context across sessions so a recurring job stays coherent over time. Work mode without those is a sports car with no road. With them, it is the closest thing in this book to handing a competent assistant a standing instruction and trusting them to carry it out.

Reach for Work when the verb in your request is an *action*, not a *question*. "Every Monday, pull last week's support tickets from the connected tool and send me a summary." "Go through my inbox, find everything from this vendor, and assemble their outstanding invoices into one document." "Take the numbers from these three spreadsheets and build me a quarterly comparison." The tell is that you are describing a task you would otherwise *do* yourself across several apps — and that you would be happy to never do by hand again. That is Work's whole reason to exist.

It helps to walk through what one of these jobs actually looks like from the inside, because "delegates and performs multi-step tasks" is abstract until you watch the steps. Take the inbox-and-invoices example. You give Work the instruction once: "Go through my connected inbox, find every message from this vendor in the last quarter, pull out any outstanding invoice amounts, and assemble them into a single document with a running total." Work does not answer that with advice. It plans the task into steps, then executes them: it queries the connected mail for messages matching the vendor, reads through what it finds, identifies the ones that contain invoice figures, extracts the numbers, sums them, and produces the document. Along the way it is making small judgments — is this message actually from the vendor or just mentioning them, is this figure an invoice or a quote — and a good first instinct is to check exactly those judgments in its early runs. The point is that Work is a sequence of real operations against real systems, not a single clever sentence, and the value is that you described the outcome once instead of clicking through all of it yourself.

The scheduling half is where this stops being a convenience and becomes infrastructure. The same job, set to run every Monday morning, becomes a standing report that arrives without you lifting a finger — last week's vendor invoices, every week, waiting in your inbox before you sit down. The mental shift here is significant: you go from *doing a task* to *defining a task and then forgetting about it*, which is exactly the relationship you have with a competent assistant who has a standing instruction. The honest flip side is that a scheduled job runs on its own judgment unattended, which is precisely why the early-run supervision below matters so much — you are about to trust it to act without you watching.

One honest note, because agentic tools deserve it: a mode that *acts* deserves more of your attention than a mode that merely *answers*. When Vibe is reading your inbox and touching your spreadsheets, you want to understand what it is connected to and what it is permitted to do, and you want to check its early jobs before you trust it to run unattended. Chapter 8 covers the setup; Part VIII covers the responsible-use posture. For now: Work is powerful precisely because it acts, and anything that acts is worth supervising while it earns your trust.

Code: The Developer's Vibe

Code is the mode built for software development — and it is a serious one, not a chat box that happens to know syntax. It reaches you three ways: as a CLI tool you run in your terminal, as a VS Code extension inside your editor, and on the web. It runs cloud sandbox sessions, so it can actually execute and iterate on code in an isolated environment rather than only suggesting it. And as of roughly mid-2026 you can kick off a coding job straight from Slack, which tells you something about where this mode is headed: toward being a developer agent you delegate to, not just an autocomplete you ride.

Code mode is for writing, editing, explaining, debugging, and running software. It understands a codebase, makes changes across files, runs them in its sandbox to see if they work, and iterates. For a developer it is less "a smarter autocomplete" and more "a junior engineer you can hand a scoped task." Part IV is built entirely around it — Chapter 12 on coding in Vibe, Chapter 13 on the CLI and editor integrations — so I will not pre-empt the detail here.

Reach for Code when you are *building or fixing software* and want the assistant working in your actual development environment, with the ability to run what it writes. If you are a developer, this is the mode you will mostly live in. If you are not, you may never open it, and that is fine — it is there when a coding task arrives.

The detail that separates Code from "a chat that knows programming" is the sandbox. When Code writes something, it can run it in an isolated cloud environment, see whether it actually works, read the error if it does not, and try again — a loop that a plain conversational assistant cannot close because it has nowhere to execute. That changes the kind of task you can delegate. Instead of "write me a function that does X" and then copying the result into your own editor to find out whether it runs, you can hand Code a scoped problem — "this test is failing, figure out why and fix it" — and let it iterate against real execution until the test passes. The analogy to a junior engineer holds precisely because of that loop: you are not asking for a snippet, you are assigning a task and getting back a verified result. Part IV is where the mechanics of the CLI, the editor extension, and the sandbox get their full treatment, so the worked detail lives there.

How to Choose, in One Breath

Strip away the detail and the choice is almost a reflex once you have it. **Ask the question: do I want Vibe to *talk* with me, to *act* for me, or to *build* for me?**

If you want it to think, write, explain, or brainstorm — to produce text in a conversation — that is **Chat.** If you want it to carry out a multi-step task across your connected apps, especially one worth repeating on a schedule — that is **Work.** If you want it to write, run, and fix software in your development environment — that is **Code.**

The most common mistake, by a wide margin, is the one I opened with: trying to get Chat to *do* a job, when Chat is built to *discuss* one. The fix is never a cleverer prompt. It is reaching for Work. The second most common mistake runs the other way — spinning up the full agentic machinery of Work for something that was only ever a question, when Chat would have answered it in a sentence. Match the mode to the verb and most of the friction in using Vibe simply disappears.

Why the Mode Shapes the Work

It is worth saying plainly why this matters more in Vibe than the analogous choices do in a single-box assistant. When you pick a mode, you are not adjusting a tone or a verbosity dial. You are choosing what *kind of thing* Vibe is allowed to be for this task: a conversational partner who hands results back to you, an agent with reach into your applications and permission to act, or a developer working in your codebase. Those are three different relationships with three different surfaces, three different risk profiles, and three different kinds of value.

That is also why the modes are not interchangeable and why "just use the smartest one" is the wrong instinct. Work is not "Chat plus." It is Chat's *opposite number* — designed to leave the conversation and touch the world, which is exactly what you do not want when you only meant to ask a question. Code is not "Work for programmers." It is its own environment with its own sandbox and its own integrations. The skill is not picking the most powerful mode. It is picking the *right* one, and the right one is defined entirely by what you are trying to accomplish.

Carry that out of this chapter and the rest of Part II falls into place. Get comfortable in Chat first, because it is the gentlest door — but keep Work in the back of your mind the whole time, because the moment you catch yourself describing a task instead of asking a question, you will know exactly which mode to reach for.

Chapter 7: Your First Conversations

Open the web app, let your coffee go cold, and hand Vibe a small real problem: "I've got a board update due Friday, here's last quarter's version — help me figure out what to change for this one." Paste the old document, hit send, and it comes back not with a vague lecture on board updates but with a specific read of *your* document — what is stale, what the new quarter will need, three questions it has to know the answers to before it can finish. In about a minute, the abstract idea of "an AI assistant" becomes a concrete thing that has just done real work on a real task in front of you. That is the whole purpose of this chapter: to get you from reading about Vibe to having a genuine first session, so the capability stops being a claim and becomes something you have actually done. Reading convinces no one. One good first conversation convinces everyone. So let us have one.

Just Start Talking

The good news is that there is almost nothing to learn before you begin. Vibe is a conversation, not a command line — you type what you want in plain language, the way you would ask a sharp, fast colleague, and it answers. There is no special syntax, no keywords to memorize, no wrong button to press. Open the web app or the mobile app, find the text box, and type a real question. The single most common mistake I see in first-timers is overthinking the *first* message, treating it like a search query they have to phrase just so. You do not. Ask the way you would ask a person.

Start in **Chat** — it is the gentlest of the three modes (Chapter 6), the one that asks nothing new of you, and the right place to build your instinct before you reach for Work or Code. Pick something you actually want, and send it. If you need a nudge, here are first messages that tend to land well — concrete, useful, and a fair test of the tool.

Something from your real work or life. "I have a presentation to a skeptical audience next week on why we should switch vendors — help me find the three strongest arguments and the weakest one I should prepare to defend." A request with real constraints shows you how Vibe handles specifics.

Something you would otherwise open a dozen tabs for. "Explain the difference between a defined-benefit and a defined-contribution pension like I'm smart but new to this." Watch how it pitches the explanation to exactly the level you asked for.

Something with your own material in it. Paste a real document — an email, a draft, a messy set of notes — and say "tighten this and tell me what's unclear." This is where Vibe stops being a trivia machine and starts being a colleague.

Send one. Read the answer. You have now had a conversation with Vibe, which is more than most write-ups will ever get you to do.

Let me make one of these fully concrete, because watching a first exchange unfold teaches more than a list of suggestions. Suppose you take the third option and paste in a real email you have been avoiding — a reply to a colleague who has asked you to take on a project you do not have time for — with the instruction "help me decline this gracefully without closing the door." A good first answer will not just spit out a single canned refusal. It will give you a draft that acknowledges the ask, declines clearly, and offers a small alternative — and it may flag a choice back to you: *do you want to suggest a later timeframe, or keep it open-ended?* That little flag is worth noticing, because it tells you Vibe caught a real ambiguity in your request rather than papering over it. You answer the question, the draft tightens, and within two exchanges you have something you would actually send. The first message got you 70 percent there; the conversation got you the rest. That ratio is the texture of using Vibe well, and you will feel it in your very first session if you let the exchange run past one turn.

Vague In, Vague Out

Now the one idea that will improve your results more than any other, and it is not a trick — it is a habit. **The quality of what you get back tracks the quality of what you put in.** A vague request earns a vague, hedge-everything answer, because Vibe is guessing at what you meant. A request with specifics earns a specific answer, because there is nothing left to guess.

Watch the difference. A weak request: *"give me some tips on writing."* Tips for whom, writing what, toward what end? Vibe will produce something generic because you handed it nothing to aim at. A strong request: *"I'm a project manager and my status emails to executives are too long and get skimmed.*

Rewrite this one to be scannable and to lead with the decision I need — here's the email." Now there is a role, a problem, a goal, and the raw material. The answer comes back sharp because the question was.

The pattern underneath generalizes to everything you will ever ask an assistant.

Say who you are or who it is for when it changes the answer — "explain this to a beginner," "I'm a clinician," "this is going to the board."

Say what you actually want — a summary, three options, a rewrite, a yes-or-no with reasons. Naming the *shape* of the output you want is half the battle.

Give it the material and the constraints — paste the text, name the budget, state the deadline, list what to avoid. Vibe cannot account for a constraint you kept in your head.

You do not have to do all three every time, and short questions are perfectly fine for simple things. But when an answer comes back mushy, nine times out of ten the fix is not a cleverer prompt — it is a more specific one. Add what you left out and ask again.

Watch the same request improve as you add each piece, because seeing the gradient is more convincing than being told about it. Start at the bottom: *"help me write a performance review."* That earns a generic template, because Vibe has no idea who, for what role, or saying what. Add the role: *"help me write a performance review for a software engineer."* Better — now the headings make sense for the job. Add the substance: *"...who shipped reliably and mentors well but avoids hard conversations with teammates."* Now it is writing about a real person rather than a category. Add the shape and the constraint: *"...keep it to half a page, lead with strengths, and frame the conflict-avoidance as a growth area, not a failing."* Now the answer is something you can edit rather than rebuild. Nothing in that sequence was a trick or a magic phrase. Each addition simply removed one more thing Vibe had to guess at, and the answer sharpened in exact proportion. That is the entire mechanism, and once you feel it you will never go back to one-line requests for anything that matters.

It is worth naming the inverse too, because over-specifying has its own small cost. You do not need to write a paragraph of context to ask what time zone Lisbon is in. The skill is matching the effort to the stakes: trivial questions deserve trivial prompts, and the moment to invest in specifics is the moment the answer would actually change depending on the details you are holding back. If adding context would not change the answer, skip it; if it would, supply it. That judgment, not prompt-craft, is what separates fast users from frustrated ones.

It Is a Conversation, So Keep Talking

Here is what first-timers most often miss: you are not allowed only one shot. Vibe holds the thread, so the most powerful move in your toolkit is the follow-up. The first answer is a draft you steer, not a verdict you accept.

So push on it, the way you would with a person who just told you something. "Shorter." "More formal." "You missed the part about the deadline — redo it with that." "Why did you recommend the first option over the second?" "Give me three more like the second one." Each turn refines, and because Vibe carries the context of everything said so far, you do not re-explain — you just nudge. I get to a good result far more often by sending a rough first request and refining it twice than by trying to compose one perfect prompt up front. Let the conversation do the work.

A few follow-up moves are worth having in your hand from the start, because they unlock things first-timers rarely think to ask for. **Ask for its reasoning.** "Why did you choose that framing?" turns a result into an explanation, and the explanation often reveals an assumption you want to correct. **Ask for alternatives.** "Give me two more versions, one bolder and one more cautious," costs you one message and gives you a spread to choose from instead of a single take you have to either accept or rewrite from scratch. **Ask it to critique its own answer.** "What's the weakest part of what you just gave me?" is a surprisingly honest move; Vibe will often surface the soft spot you were already half-sensing. **Correct the premise, not just the output.** If the whole direction is wrong, do not nudge the wording — say "you've misread what I'm after; I don't want a summary, I want a decision recommendation," and let it reorient. These are not advanced techniques. They are just the natural things you would say to a capable colleague, and the only reason to list them is that the blank box does not advertise that you are allowed to.

There is also a quiet decision hidden in the follow-up loop: when to keep steering a tired thread versus when to start fresh. If you have refined something five or six times and it keeps drifting back toward a version you did not want, the conversation has usually accumulated enough conflicting instructions that a clean restart — pasting in your best draft and stating clearly what you want — beats one more nudge. You will not need this in your first session, but it is good to know the escape hatch exists: a stuck conversation is not a failure, it is just a signal to begin again with what you have learned.

Notice When You Want More Than an Answer

Everything so far is Chat, and Chat will carry most of your first session. But there is a moment worth learning to recognize, because it is the doorway to what makes Vibe distinctive, and you may hit it sooner than you expect.

It is the moment you stop asking a *question* and start describing a *task*. Watch for the verb. "Explain how I'd reconcile these two spreadsheets" is a question, and Chat answers it. "Go reconcile these two spreadsheets and give me the differences" is a task — and that is **Work** mode (Chapter 6), the agentic mode that actually goes and does things across your connected apps rather than just telling you how. You do not need Work in your first session, and I would not start there; Chat is the gentler ground. But the instant you catch yourself wishing Vibe would *do* the thing instead of *describe* it, you have found the edge of Chat and the reason Work exists. File that feeling away. It is the single most useful instinct you can develop in your first week, and the next chapter — on the Libraries, Connectors, and Memory that power Work — is where it pays off.

A Few Honest First-Session Notes

So your expectations are calibrated and not just inflated: Vibe is fast, fluent, and broadly capable, and it can also be confidently wrong. A smooth, assured answer is not automatically a correct one — fluency is not accuracy, and that is true of every model in the field, Mistral's included. For anything that matters — a number you will act on, a fact you will repeat, a claim you will stake something on — verify it, especially when the answer leans on a web search you can go check yourself. This is not a knock on Vibe; it is the basic posture for using any assistant well, and it is cheap insurance.

You will also notice that Vibe answers easy things almost instantly and takes a beat longer on hard ones, especially when you have asked for the deeper reasoning your plan unlocks (Chapter 5). Do not be surprised by the pause on a genuinely hard question; that is Vibe thinking, and it is usually time well spent.

And one Mistral-specific note worth a moment's pride if it matters to you: where your data goes is part of what you are choosing when you choose Mistral at all. For an individual on the Vibe app that story is mostly abstract — you are using a hosted assistant like any other, and the privacy terms are what govern (Chapter 16). The fuller sovereignty proposition lives at the business and enterprise level (Part V). It is worth knowing which half of Mistral you are touching, so your expectations stay honest from the first session on.

Go Have the Conversation

If you take one thing from this chapter, let it be that there is no preparation required and no way to do it wrong — there is just the conversation, which gets better the more specific you are and the more you follow up. Start in Chat, pick a real question, ask it plainly, paste in some of your own material, and refine the answer a couple of times to feel how steering works. And keep one ear open for the moment you stop asking and start delegating, because that is the edge where Vibe becomes something more than a chatbot. Most of the rest of this book is refinement on top of that loop: ask, read, refine, verify what matters — and, when the work is a task rather than a question, hand it to Work. You already know how to begin. Open the box and say something real.

Chapter 8: Libraries, Connectors, and Memory

The moment Vibe stopped feeling like a clever stranger and started feeling like a colleague was the day it remembered something I had not repeated. I had spent a session a week earlier explaining how our reports were structured — the sections, the order, the tone the board expected — and when I came back to ask for a new one, I braced to explain it all again. I did not have to. Vibe already knew, picked up where we had left off, and produced something shaped like *our* reports rather than a generic one. Nothing about that is magic, and that is exactly the point. It was three features doing quiet work underneath: a Library holding the knowledge, Memory carrying the context across the week, and — the time before, when it had pulled real numbers — a Connector giving it somewhere to reach. These three are what turn Vibe from a smart conversation into a persistent, context-aware assistant, and they are what make the agentic Work mode of Chapter 6 actually possible. This chapter is about all three: what they are, how to set them up, and why they matter more than any single clever answer.

The Problem They Solve

Start with the limitation they exist to fix. A plain assistant, however capable, is an amnesiac with no hands. Each conversation starts from nothing — it does not remember last week, does not know your documents, and cannot touch anything outside the chat box. You can paste context in every time and read its suggestions out and act on them yourself, but you are the memory, the filing cabinet, and the hands. That works, and for a lot of Chat-mode work it is fine. But it is also the ceiling that separates a chatbot from an agent.

Libraries, Connectors, and Memory are how Vibe breaks that ceiling. Libraries give it knowledge to draw on. Connectors give it reach into your applications. Memory gives it continuity over time. Together they are the difference between an assistant you have to re-brief every session and one that knows your context, your tools, and your history — which is precisely what an agent in Work mode needs to act on your behalf without you holding its hand through every step.

Libraries: Saved Knowledge Vibe Can Draw On

A Library is a place to store content and knowledge that Vibe can reference — documents, reference material, the standing facts about how your work works. Instead of pasting the same background into every conversation, you put it in a Library once, and Vibe can draw on it whenever it is relevant.

Think of what belongs there. Your style guide. The template your reports follow. The reference documents your team keeps returning to. The product specs, the policy language, the boilerplate, the institutional knowledge that you would otherwise re-explain over and over. Put it in a Library and it becomes part of what Vibe knows about *your* world, not just the general world it was trained on.

To set one up, you create a Library and add content to it — uploading documents or saving material from your conversations. The practical advice is to start small and specific rather than dumping everything you own into one undifferentiated pile. A focused Library on a single topic or workflow gives Vibe clean, relevant knowledge to pull from. A vast junk drawer gives it noise. Build the Libraries that match the work you actually repeat, and let them grow as you find yourself re-explaining the same things.

It helps to see what a well-built Library does in practice, because the payoff is concrete. Suppose you keep a Library holding your team's report template, your house style guide, and a one-page "about this project" brief. When you later ask, "draft me a status report on the migration project," Vibe is no longer guessing at your conventions — it pulls the structure from the template, the tone from the style guide, and the background from the brief, and what comes back is shaped like *your* organization's work rather than a generic report. The first time this happens it feels like the assistant suddenly "knows" your job. It does not, in any deep sense; it simply has clean, relevant material to draw on, supplied once instead of pasted every time. That is the entire trick, and it is why the quality of a Library matters more than its size.

Which is also why the start-small advice is not just tidiness. A Library is reference material Vibe will *lean on*, so noise in it becomes noise in your answers. A Library that holds your current style guide plus three good example documents produces sharp, on-brand drafts. The same Library after a year of dumping every file you touched — superseded templates, abandoned drafts, a contract from a project that ended — produces muddier ones, because Vibe cannot reliably tell the live reference from the dead weight. The maintenance habit that pays off is the same one you would apply to a shared team drive: prune the stale, keep the current, and let each Library stay focused on a single workflow rather than becoming a junk drawer that happens to be searchable.

On a Team plan, Libraries can be shared (Chapter 5), which is where they earn their keep collectively — the whole group draws on the same institutional knowledge instead of each person maintaining a private copy in their head. That shared-knowledge layer is one of the strongest reasons a working group moves from individual Pro seats to a Team workspace. The collective version also changes the maintenance stakes: a shared Library that drifts stale steers everyone's answers wrong, not just yours, so a team benefits from naming someone who owns keeping it current the way someone owns the shared drive.

Connectors: Giving Vibe Somewhere to Act

A Connector links Vibe to one of your apps or tools, so that it can read from — and act on — that application. This is the feature that makes Work mode more than a thought experiment. A Connector is the road; without it, the agentic engine has nowhere to drive.

This is what lets Work mode do the things Chapter 6 described: scan an inbox, because you have connected your mail; pull figures out of spreadsheets, because you have connected the place they live; gather information from across several tools and assemble it into a report, because each of those tools is connected. The Connector is the bridge between Vibe's reasoning and your actual systems. Connect nothing, and Work mode can only think; connect your apps, and it can act.

You set a Connector up by authorizing Vibe to access a given application, granting it the permissions the task requires. And here is the place to slow down, because a Connector is exactly the kind of feature that deserves a careful hand. You are giving an agent reach into a real system — your mail, your files, your business data. That is genuinely powerful, and power that acts deserves more scrutiny than power that merely answers. Three habits are worth forming from the start.

Connect deliberately, not reflexively. Add the Connectors a workflow actually needs, rather than wiring up everything because you can. Each connection is a door; open the ones you will use and leave the rest shut.

Understand the permissions you grant. Read what access a Connector is asking for. Whether Vibe can only read, or can also write and send, changes the stakes considerably, and you want to know which you have agreed to.

Watch the early jobs before you trust the standing ones. When a Connector lets Work act on your behalf, supervise its first runs and confirm it does what you meant before you let it run unattended on a schedule. Part VIII treats the responsible-use posture in full; the short version is that anything with hands deserves a few supervised rehearsals.

The read-versus-write distinction in that second habit deserves dwelling on, because it is the single most consequential thing about a Connector and the easiest to wave past during setup. A read-only connection lets Vibe *see* — scan your inbox, pull figures from a sheet, find a file. The worst a mistake there can do is give you a wrong summary, which you catch by reading the source. A write-or-send connection lets Vibe *change the world* — send the email, edit the document, post the message. Now a mistake leaves the building before you see it. These are genuinely different risk levels wearing the same word "Connector," and the authorization screen is where you choose which one you are agreeing to. The discipline that follows from this is simple: grant read access freely where a workflow needs visibility, and grant write or send access deliberately, to the specific workflows that truly require it, with your eyes open to the fact that you have just handed an agent the ability to act in your name.

A short worked example makes the early-supervision habit concrete. Say you connect your mail and ask Work to "reply to routine scheduling emails by proposing times from my calendar." The sane way to bring that online is in stages. First, run it in a mode where it *drafts* the replies and shows them to you rather than sending — you read several, confirm it is proposing sensible times and not double-booking you, and only then do you let it send on its own. You are not distrusting the tool for its own sake; you are doing the exact thing you would do with a new assistant on their first week, which is to check the early work before you stop checking. Once a handful of runs come back clean, the supervision can relax. The mistake is skipping the rehearsal and discovering on day one that "routine" included an email it should not have touched.

None of that is a reason for hesitation — Connectors are where Vibe becomes genuinely useful as an agent. It is a reason for the ordinary care you would extend to any new hire you handed the keys.

Memory: Continuity Across Sessions

Memory is Vibe's ability to remember context across conversations, so that what you established last week is still there this week. It is why my report request picked up where the previous one left off, and it is what spares you the tax of re-briefing the assistant every single time.

What Memory captures is the durable context of working with you — preferences you have expressed, facts about your situation, the way you like things done — carried forward so each session builds on the last instead of starting from zero. For a one-off question this is invisible. For an ongoing working relationship, and especially for a recurring Work-mode job that runs on a schedule, it is essential: a recurring task stays coherent over time precisely because Vibe remembers what the task is and how you want it done.

It is worth being precise about the difference between Memory and the thread of a single conversation, because they are easy to conflate and they fail differently. Within one conversation, Vibe already remembers everything said so far — that is just the thread, and it ends when the conversation does. Memory is the layer *above* that: the durable facts and preferences carried from one conversation into the next, days or weeks apart. So "you forgot what I said two messages ago" is a thread problem (usually a very long conversation straining its limits), while "you forgot the preference I set up last month" is a Memory problem. Knowing which one you are looking at tells you whether the fix is to start a fresh, focused conversation or to go check what Memory actually has stored.

Because Memory persists, it is worth knowing you have a hand on it. You can see and manage what Vibe remembers — correcting something it has gotten wrong, removing context that has gone stale, keeping its picture of you accurate. Treat that as routine hygiene rather than an afterthought: a Memory that has quietly drifted out of date will steer answers subtly wrong, and the fix is simply to tidy it. The failure mode to watch for is the obsolete fact that quietly keeps shaping answers — Memory still thinks you are on the old project, or still applies a preference you have outgrown, and so every response leans a few degrees off true without ever being obviously wrong. The remedy is the same light habit you would apply to any standing record: glance at what Vibe remembers now and then, correct what has gone stale, and remove what no longer applies. Knowing the controls are there, and using them now and then, keeps Memory an asset rather than a slowly accumulating liability.

How the Three Work Together

The reason this chapter treats them as one subject is that their real power is combined. Each is useful alone; together they are what makes agentic Work possible.

Picture a standing job: every Monday, Vibe assembles your weekly status report. **Connectors** let it reach into your mail and your spreadsheets to gather the week's raw material. A **Library** holds the template and the standards, so the report comes out shaped the way your organization expects. **Memory** carries forward your preferences and the shape of the task, so the third Monday's report is as coherent as the first. Pull any one of the three and the job degrades — no Connectors and it has nothing to gather, no Library and it has no shape to fill, no Memory and it forgets how you wanted it done. Together, they let you hand Vibe a standing instruction and trust it to carry it out week after week. That is the whole agentic proposition, and these three features are its foundation.

Where to Start

Do not try to build all of this at once. The right sequence is to let the work pull the features in. Start in Chat, as Chapter 7 suggested, and notice the friction. The first time you find yourself pasting the same background for the third time, build a **Library** for it. The first time you wish Vibe would *act* on a real system instead of describing the action, add the **Connector** that gives it that reach — carefully, with the permissions understood. And let **Memory** accumulate naturally, checking on it now and then to keep it honest.

Build them in that order — knowledge first, reach second, continuity throughout — and you will find Vibe quietly becoming less of a clever stranger and more of a colleague who knows your work. That shift, from answering to acting, from session-by-session to persistent, is what the rest of this book builds on. With these three in place, you are no longer just using an assistant. You have started to delegate.

Part III: Vibe Work

Audience: everyone. This is Vibe as an agent — delegating multi-step work, not just answering questions.

"The shift from a chatbot to a work agent is the shift from asking 'what's the answer' to asking 'can you just handle this' — and then reading what comes back."

— A working principle

Chapter 9: Writing and Documents

I had a forty-page vendor contract to get through on a day I did not have an hour to give it, and a meeting in twenty minutes where someone would ask me what was in it. I dropped the PDF into Vibe and asked for the obligations that fell on us, the payment terms, and anything that looked unusual for a contract of this kind. What came back was a tight, organized summary with the clauses named — and one flag I would have missed, an auto-renewal buried in a section I would not have read closely. I walked into the meeting able to speak to the document. Then, afterward, I went back and read the three clauses Vibe had flagged, in the actual contract, because a summary is a map and I was about to sign the territory. The summary saved me twenty minutes. The reading is what kept me honest. That gap between the two is what this chapter is about.

This chapter is about using Vibe — mostly Work mode, sometimes Chat — for the writing that fills a working life: drafting, editing, summarizing, and working with documents you upload. Vibe is good at all of it. It is not a replacement for reading your own important documents, and the discipline of checking what it hands back matters more the higher the stakes. We will take the capabilities one at a time and keep the review habit in view throughout.

Drafting

The blank page is where an assistant earns its keep fastest, because the hardest part of writing is often just having something to react to. Describe what you need — an email declining a meeting without burning the relationship, a project update for a skeptical stakeholder, a first pass at a job description, the bones of a blog post — and Vibe produces a draft you can shape. The draft is rarely the final thing, and it should not be. It is a starting point that turns the work from *generate from nothing* into *edit something*, which is a far easier kind of work and a far faster one.

The quality of what you get back tracks closely with the quality of what you put in. "Write a follow-up email" gets you a generic follow-up email. "Write a short follow-up to a client who went quiet after we sent a proposal three weeks ago, warm but not needy, with a light prompt to reconnect" gets you something you can almost send. **The specifics you provide — audience, tone, length,**

purpose, what to avoid — are the difference between a draft you rewrite and a draft you polish. This is the same lesson as Chapter 7, applied to writing: tell it what you actually want, including the constraints, and it meets you most of the way.

There is a drafting move that earns its place once you trust it: asking for several drafts at once rather than one. "Give me three versions of this announcement — one warm, one brisk, one formal" costs you the same single request and hands back a spread you can react to. Choosing between three concrete options is far easier than conjuring one from nothing or judging a lone draft against an idea you have not articulated, and you will often find the version you want is a splice — the opening of the brisk one, the closing of the warm one. Vibe will happily combine them on request. The blank page rarely survives contact with three drafts to argue with.

Where drafting pays off most is the writing you find tedious rather than the writing you find meaningful. The routine status update, the polite decline, the meeting agenda, the first scaffold of a longer piece — this is the work you would rather not start, and handing the cold open to Vibe so you can step in as editor is a genuine and honest use of the tool. The writing that carries your voice and your judgment, you will still want to drive yourself. Vibe is leverage on the parts you would skip, not a ghostwriter for the parts that are the point.

The honest failure mode of drafting is the one that looks like success: a fluent, confident draft that is subtly wrong about a fact, a name, or a commitment. Vibe will write "as we agreed last week" or cite a figure with complete assurance whether or not it has any basis for either, because producing plausible prose is exactly what it is good at and verifying the underlying claim is not the same skill. So the draft you hand off cleanest is the one whose facts you supply — paste the real numbers, name the real dates, state the real commitment — rather than trusting it to remember or infer them. Treat a generated draft as confident about its style and unreliable about its facts, and read it for the second even when the first is carrying you along.

Editing

Drafting gets the attention, but editing may be the more valuable everyday use, because most professionals write more than they realize and edit worse than they think. Paste in something you wrote and ask Vibe to tighten it, fix the grammar, make it warmer or more formal, cut it to half the length, or simply tell you where it drags. The fresh-eyes effect is real here the same way it is for code: the model has not been staring at your sentence for ten minutes and does not love your favorite phrase the way you do.

Three editing moves I reach for constantly. **Tightening** — "make this half as long without losing anything important" is a request Vibe handles well, and concision is the edit most writing needs most. **Tone-shifting** — the same content recast for a different reader, the blunt internal note softened for a client, the casual draft made presentable. **Diagnosis** — not "fix this" but "what is weak about this," which turns the assistant into a critic you can argue with rather than a rewriter that flattens your voice into its own.

The diagnostic move deserves a moment of its own, because it is the one most people never think to ask for and the one that improves their writing the most. Instead of "rewrite this," try "read this and tell me, in order, the three things that weaken it most." What comes back is a critique you can act on with your own hands: this paragraph buries the ask, this sentence hedges, this section repeats the one before it. You keep your phrasing and your voice; you fix the structural problems Vibe is genuinely good at spotting. Over time this is also the better teacher — wholesale rewrites give you a fish, diagnoses teach you to see the weaknesses yourself, and after a few dozen you start catching them before you ask.

That last point carries a warning. Vibe has a house style, and if you let it rewrite freely it will sand your prose toward a competent, slightly anonymous middle. You can feel it happen: the version it hands back is cleaner and somehow less *yours*, the specific turn of phrase you liked smoothed into the kind of sentence any assistant would produce. For a routine internal memo that flattening costs you nothing. For anything where your voice is part of the message — a note to your team, a piece with your name on it, writing meant to sound like a person — it costs you the thing that made the writing worth reading. For that writing, prefer the diagnostic mode — ask what is wrong and fix it yourself — over wholesale rewriting. The goal is your writing, made better, not the model's writing wearing your name.

Summarizing and Working with Uploaded Documents

This is where Vibe Work earns the most time, and where the review discipline matters most. You can upload documents — reports, contracts, research papers, long email threads, slide decks, transcripts — and ask Vibe to summarize them, pull specific information, answer questions about them, or compare them against each other. The forty-page contract is the canonical case: minutes of reading compressed into a summary you can act on, with the option to drill into any part.

The capabilities worth knowing. **Summarize** a long document to its essentials, at whatever altitude you ask for — one paragraph or one page. **Extract** specific things — every deadline, all the dollar figures, the action items, the names — which turns an unstructured document into the list you actually needed. **Answer questions** against the document — "does this contract let either side terminate early?" — so you can interrogate it rather than read it linearly. **Compare** two or more documents, which is tedious by hand and fast for the model: what changed between these contract versions, how do these two proposals differ.

And the honest limits, because uploaded-document work has failure modes that look exactly like success. A summary can be confidently incomplete — it can miss the one clause that matters precisely because it was buried and unusual, the very thing you most needed it to catch. The model can misread a number, attribute a statement to the wrong party, or smooth over a nuance that changes the meaning. On a long document it can lose track of something in the middle. None of this announces itself; a wrong summary reads as fluently as a right one.

A worked example shows how to get the leverage without the exposure. Say you upload a thirty-page research report and need to brief your team on it tomorrow. The efficient path is not "summarize this" and done. It is a sequence: first, "give me a one-paragraph summary and then a one-line takeaway per section," which gives you the shape; then, "what are the three claims this report makes that would most change our decision if they're true?"; then, for each of those three, "show me where in the document that claim is made" — and *that* is where you go read, in the actual report, the specific passages your decision will rest on. You have used Vibe to navigate thirty pages down to the three paragraphs that matter, and you have read those three with your own eyes. That is the whole discipline in motion: the summary is the index, not the verdict, and the reading is targeted rather than abandoned.

So the discipline, which is the writing-and-documents version of the read-the-code rule that runs through this book: **for any document that carries real consequence, the summary is a guide to your reading, not a substitute for it.** Let Vibe tell you where to look and what to look for — that is enormous leverage, and it is the legitimate win. Then read the parts that matter in the actual document, especially anything you are about to rely on, sign, send, or decide from. The summary tells you the contract has an auto-renewal clause; you read the clause before you sign. For low-stakes documents — the long thread you just need the gist of — the summary alone is fine. Calibrate the checking to the cost of being wrong, and never skip it where being wrong is expensive.

Weaving in Libraries and Connectors

Writing and documents get more powerful when Vibe can reach your context, and that is what Libraries, Connectors, and Memory (Chapter 8) add to this work. The plain version of document handling is one-off: you upload a file, you ask about it, you move on. The connected version is where it becomes part of how you work.

With **Connectors**, Vibe can reach documents where they already live — a file in your connected drive, a thread in your connected email, content in a connected workspace — so you are not forever uploading the same files by hand. "Summarize the latest version of the proposal in our shared drive" becomes a thing you can ask, rather than a thing you first have to go find and download. This is also the seam where Chapter 9 hands off to the agentic work of Part III: once Vibe can reach your documents and apps, summarizing one file is the small version of a larger pattern where it scans, pulls, and routes information across them on your behalf (Chapters 10 and 11).

With **Libraries**, you can keep reference material on hand — your style guide, your templates, background on a recurring project, the boilerplate you reuse — so Vibe writes with your context rather than from a cold start. A draft produced against your actual style guide and past examples is closer to usable than a generic one, because the model is no longer guessing at conventions it can simply be shown. And with **Memory**, the preferences you have established carry forward, so you are not restating "keep it concise, no exclamation points, sign off as the team" every time.

One caution on connected work, and it is the same caution as everywhere else in this part. The more Vibe reaches into your real documents and apps on its own, the more its mistakes touch real things rather than a sandbox. A summary you read before acting is low-risk. An agent that reads a document and then sends something based on it is a different exposure, and Part III is where we treat that seriously. For the writing and document work in this chapter, the rule stays simple: use the connections for reach and context, keep your eyes on anything consequential before it goes out the door, and remember that the assistant drafts and summarizes — you decide and you sign.

That distinction, between a tool that drafts and a person who decides, only grows sharper when Vibe stops shaping your words and starts telling you what is true — when it goes out to the live web to research a question, and the stakes of checking the source climb again.

Chapter 10: Research and the Web

Web research is a real superpower, and checking the sources is the price of using it safely. Hold those two facts together and you have the whole shape of this chapter. Picture the common case: you need a quick read on a regulation that changed recently, the kind of thing where your own knowledge is a year stale and getting it wrong would be embarrassing in the meeting you are prepping for. Vibe goes out to the web, pulls together a clear summary, and — this is the part that matters — gives you the sources it drew from. Read the summary, then open two of the cited pages, and you may find that one of them says something slightly different from how Vibe characterized it: not wrong exactly, but flattened in a way that would have led you to overstate the change. The research saves an hour of hunting. The two clicks to the sources save you from saying something confidently incorrect.

This chapter is about Vibe's ability to search the web and produce researched, sourced answers — including deeper, report-style research on a question — and about the discipline that has to ride alongside it. The capability is genuinely useful and changes what you can ask. It also introduces a failure mode that is more dangerous than a model simply not knowing something, because a confident, well-formatted, *sourced-looking* answer is exactly the kind of wrong that slips past you.

What Web Research in Vibe Does

A model frozen at its training cutoff has the same blind spot every such model has: it does not know what happened after that date, and it does not know what it does not know, so it will sometimes answer a question about the present as if its old knowledge still held. Vibe's web research closes that gap. When a question turns on current information — recent news, a changed regulation, a product that shipped this quarter, the latest figures on something — Vibe can go out to the live web, read what it finds, and answer from current sources rather than from memory alone.

Two registers are worth distinguishing. The lighter one is a **quick lookup**: a question that needs a current fact or two, where Vibe searches, finds, and answers in the flow of conversation. The heavier one is **deep research** — a report-style mode where, given a substantial question, Vibe runs an extended, multi-source investigation and hands back a synthesized, organized answer with its sources cited. The first is for "what's the current rate on this"; the second is for "give me a briefing on the state of this market, with the major players and recent moves." Deep research is slower by design, because it is doing more, and it is the mode to reach for when you want a researched document rather than a fast answer.

It helps to feel the difference in practice. Suppose you are pricing a small business loan and you ask, "what is the current base rate the European Central Bank is using?" That is a quick lookup: one fact, current as of today, a single authoritative source, answered in the flow of the conversation in a few seconds. Now suppose instead you ask, "give me a briefing on how small-business lending conditions have shifted in the eurozone over the past year, the main drivers, and which way analysts expect rates to move." That is deep research. Vibe will spend longer, consult several sources, and hand back something with structure — sections, a few competing views, citations against each claim. The tell that you want deep research rather than a quick lookup is that the question has *shape*: it asks for synthesis across sources, not retrieval of one. When you find yourself wanting "the landscape" or "the state of," reach for the heavier mode; when you want "the number" or "the date," the lighter one is faster and just as good.

The defining virtue of both, the thing that makes them trustworthy enough to use, is **citation**. Vibe shows you where its claims came from. This is not a cosmetic feature. It is the mechanism that makes the output checkable, and a researched answer without sources you can inspect is one you have no way to verify and therefore should not lean on for anything that matters.

How to Ask So You Get Something Checkable

The quality of a researched answer depends more on how you frame the question than most people expect, and a few habits in the asking save a great deal of checking later. The first is to **say when recency matters**. If the answer turns on the current state of something, say so — "as of now," "the latest figures," "what changed this year" — because that signals Vibe to go to the live web rather than answer from memory. A question phrased as if it were timeless ("what is the corporate tax rate in Ireland") invites a stale answer; the same question phrased with a recency cue ("what is the current corporate tax rate in Ireland, and has it changed recently") pushes the assistant toward sources from now.

The second habit is to **ask for the sources explicitly when you care**, and to ask for a particular *kind* of source when the kind matters. "Summarize the new packaging regulation and cite the official text" is a better prompt than "tell me about the new packaging regulation," because it points Vibe at the primary document rather than at whatever secondary coverage ranks highest. When the truth lives in a primary source — a statute, a filing, a manufacturer's own spec sheet, a published standard — naming that you want the primary source is the single most effective thing you can do to make the answer worth trusting.

The third is to **scope the question to what you can verify**. A research request so broad that it would take a human a week to confirm is one whose citations you will never actually check, which means you will end up trusting it blind — exactly the trap this chapter is about. A tighter question produces a shorter chain of claims, and a shorter chain is one you can actually walk. If you need the broad sweep, get the briefing, then pull the two or three load-bearing claims out of it and verify those specifically, rather than pretending you have checked the whole document.

Why Sourced Answers Change What You Can Ask

With a frozen assistant you learn, without quite noticing, to stop asking about anything recent — the way you stop asking a reference book about this morning. You confine your questions to things you assume it knew at training time, and you go elsewhere for the present. Web research lifts that ceiling. You can ask about the current state of a fast-moving field, the recent developments on a story, the latest on a regulation or a release, and get an answer grounded in sources from now rather than from whenever the model was trained.

That changes the kind of work you bring to it. Background research that used to mean a dozen open tabs and an hour of reading becomes a question you can ask, with the assistant doing the gathering and the first-pass synthesis. The competitive-landscape scan, the literature-style summary of a topic, the "catch me up on what's happened here" — these become Vibe tasks. The leverage is real and worth taking. But — and this is the through-line of the chapter — the leverage is in the *gathering and organizing*, not in the *truth-certifying*. Vibe can find and arrange the sources far faster than you can. It cannot guarantee it read them correctly, and that judgment stays yours.

The Verify-Your-Sources Discipline

Here is the honest accounting, and it deserves its own section because the danger is specific. A large language model can be confidently, fluently wrong. It can state something false with exactly the same poise it uses for something true, and the model's confidence tells you nothing about whether the claim is correct. Web research changes the texture of this risk without removing it, and in one way it sharpens it.

Search reduces some errors and introduces others. Pulling from live sources makes Vibe less likely to be simply out of date, which is a genuine gain. But the live web is not a clean repository of truth. It contains errors, outdated pages, marketing dressed as fact, partisan spin, and the occasional outright fabrication, and Vibe is reading that mixed bag. A confident answer drawn from a wrong or biased source is still a wrong answer, now wearing the credibility of having been "researched."

The model can misread what it found. Even when the underlying sources are good, Vibe can characterize them imperfectly — overstate a finding, blur a qualifier, attribute a claim to the wrong source, or summarize a nuanced page into a cleaner statement than the page actually supports. This is the failure in the regulation case: the source was fine; the summary of it was subtly off. You only catch this by reading the source.

A citation is a pointer, not a proof. This is the subtle trap and the most important sentence in the chapter. The presence of a source link makes an answer *feel* verified, but the citation only tells you where Vibe got something — not that Vibe represented it correctly, not that the source itself is right, not that the source even says what the answer claims. An unchecked citation buys you false confidence, which is worse than no confidence, because it stops you looking.

So the discipline, plainly stated: **for anything that matters, follow the citations and read the sources yourself.** Concretely, this means a few habits. Open the cited sources for any claim you are going to rely on, and confirm the source actually says what Vibe says it says. Weigh the sources — a primary document or a reputable outlet is worth more than an anonymous page, and Vibe will not always rank them for you. Be most skeptical exactly where it would be most costly to be wrong, and be most skeptical of claims that are surprising, precise, or convenient, because those are the ones most worth a second look. For low-stakes curiosity, the sourced summary alone is fine; for a decision, a published claim, or anything with consequences, the sources are not optional reading.

A Worked Spot-Check

It is worth slowing the regulation example down to see what a real spot-check looks like, because "read the sources" is easy to nod at and easy to skip. Say you asked Vibe to summarize a change to a data-protection rule, and it told you, with a citation, that "the new rule requires breach notification within 48 hours." You are about to put that line in a slide for your compliance team. The check takes about ninety seconds. You open the cited page. You search it for the word "notification" or the number "48." And one of three things happens, each instructive.

Sometimes the page says exactly that, in those words, and you are done — the citation held, the summary was faithful, and your ninety seconds bought you certainty. Sometimes the page says "72 hours," and you have just caught a fabricated or misremembered specific that would have been quietly wrong in front of your team; you fix the slide and move on, grateful. And sometimes — the most interesting case — the page says "without undue delay, and in any event within 72 hours where feasible," and you realize Vibe flattened a hedged, conditional requirement into a hard number. The source was not wrong and Vibe was not exactly wrong, but the *summary* lost the qualifier that does the real work, and presenting the flattened version would have made you sound more certain than the rule actually is. That third case is the one no amount of model improvement reliably prevents and the one only reading catches. The numbers in this example are illustrative — the point is the motion: open, search, compare the claim to the words on the page.

When Research Is Most Likely to Fail You

Knowing where the failures cluster lets you aim your skepticism. Research is most fragile on **fast-moving topics**, where the live web is full of outdated pages that have not caught up to a recent change, and Vibe may surface the old consensus alongside or instead of the new reality. It is fragile on **contested topics**, where sources disagree by design and a synthesized "answer" can paper over a genuine dispute, presenting one side's framing as settled fact. It is fragile on **precise quantitative claims** — exact figures, dates, dollar amounts, percentages — which are both the easiest things to get subtly wrong and the easiest to check, so they deserve a disproportionate share of your verification. And it is fragile wherever the **best source is paywalled or hard to reach**, because Vibe may fall back on a secondary summary of the real source rather than the source itself, adding a layer of someone else's interpretation between you and the truth. None of these makes research unusable. Each tells you where the two clicks are most worth spending.

This is the research version of the read-the-code rule from the coding chapters and the read-the-document rule from Chapter 9. The same principle wears different clothes throughout this book: the assistant gives you leverage, and you keep the judgment. Web research lets you gather and synthesize current information at a speed you could not match by hand. It does not transfer to the machine the responsibility for whether the information is true. That stays with you, and the citations are the tool the system hands you to discharge it — useful only if you actually use them.

Where Research Fits in the Work

Web research is not a standalone island; it threads into the rest of what Vibe does. A researched answer can feed straight into the writing of Chapter 9 — research a topic, then draft from it — and the sources can be saved into a Library (Chapter 8) so a recurring research task starts from your accumulated context rather than cold. And research is a building block of the agentic work ahead: a Work-mode agent that "sends me a weekly summary of developments in our space" is, underneath, web research run on a schedule and routed into a report (Chapter 11). The verify-your-sources discipline travels with it into those settings, and arguably matters more there, because an automated research job runs without you watching each result — which means the habit of spot-checking what it produces has to be built into how you set it up.

The realistic posture is the one the regulation story opened with. Treat Vibe's research as a fast, capable research assistant who gathers the sources and drafts the synthesis, and treat yourself as the editor who reads the sources before the findings go anywhere that counts. Used that way, web research is one of the most time-saving things in the tool. Used credulously, the citations become decoration on confident mistakes. The difference is two clicks, and those two clicks are the job.

Chapter 11: Agents and the No-Code Builder

Every Monday morning for years I did the same small ritual: open my inbox, scroll back through the weekend, and pull out the handful of messages that actually needed me before the week started. It took fifteen minutes and a cup of coffee, and it was exactly the kind of low judgment, high tedium work a machine should do. So I built an agent in Vibe to do it — no code, just a description of what I wanted: each Monday at eight, scan the weekend's inbox, summarize what came in, and flag anything that looked time-sensitive. It showed me its plan before it ran the first time. I read the plan, adjusted one thing, and let it go. Now I get the summary before I open my coffee. But I read the agent's plan that first morning, and I still read its summaries rather than acting on them blind — because an agent that touches my real inbox is doing real things, and "it ran on its own" is not the same as "it was right."

This chapter is about Vibe Work's agents and the no-code builder that lets you create them: building tools that do recurring, multi-step work across your connected apps without writing any code. This is the most genuinely new thing in Vibe relative to a plain chatbot, and it is powerful. It is also the place where the review discipline that runs through this book stops being a nicety and becomes structural, because an agent acts when you are not watching.

What a Work-Mode Agent Is

In the earlier chapters, Vibe responded to you. You asked, it answered; you uploaded, it summarized; you requested research, it researched. An agent inverts the rhythm. Instead of doing one thing when you ask, an agent does a defined sequence of things, on a trigger or a schedule, without you driving each step. It is the difference between a colleague who answers your questions and a colleague to whom you have delegated a standing task.

The mechanism that makes this possible is **Connectors** (Chapter 8), the bridges that let Vibe reach your actual apps — your email, your calendar, your drive, your spreadsheets, your messaging and project tools. An agent is, in essence, a plan that uses those connectors to move information between your apps and act on it.

Scan an inbox, pull data from a spreadsheet, assemble a report, route the result somewhere, do it again next week — these are the moves agents string together. The Monday inbox summary is a small one. The pattern scales to genuinely useful multi-step work.

What Agents Can Actually Do

Concretely, and drawn from the kinds of recurring work that fill an organization, Work-mode agents handle jobs like these.

Inbox and message triage. Scan email or a messaging channel, summarize what came in, surface what needs attention, draft replies for your review. The recurring "catch me up and tell me what's urgent" task.

Pulling and assembling data. Reach into a connected spreadsheet or database, pull the relevant numbers, and build a report or a summary from them — the weekly metrics roundup, the status digest, the figures someone always asks for on Friday.

Routing and handoff. Take an output and send it onward — post the summary to a channel, file the report in a drive, update a tracker — so the result lands where it is needed rather than sitting in the chat.

Scheduled recurring jobs. The same task on a cadence: a Monday morning summary, a daily standup digest, a weekly status report, a recurring research briefing (Chapter 10). You set the schedule once; the agent shows up on it.

To make these less abstract, here are three agents built around real recurring work, each a little more ambitious than the last. The first is a **read-only digest**: every weekday at 7 a.m., scan a shared project channel and a particular email label, summarize what came in overnight, and post the summary back to you. It touches nothing it cannot un-touch — its entire output is a summary you read or ignore — which makes it the safest possible first agent and the one most people should build before any other. The second adds **assembly across apps**: every Friday at noon, pull the week's closed support tickets from a connected spreadsheet, count them by category, draft a short status note, and file it as a document in a shared drive. Now the agent is moving data between two apps and producing a record other people will see, which raises the stakes a notch — a miscount or a misfiled note has an audience. The third adds **action with consequence**: when a new entry lands in an intake form, draft a personalized acknowledgement email and *queue it for your approval* before anything sends. That last clause is the whole difference between a useful helper and a liability, and it is the design choice the rest of this chapter is about.

Notice the progression. Each agent does more, reaches further into your real systems, and carries a larger blast radius if it goes wrong. The capability grows; so does the care the setup demands. The right way to adopt agents is to walk that ladder deliberately rather than jumping to the top rung because the builder makes it easy.

The honest frame on capability is that agents are very good at the *connective, repetitive, rule-shaped* work — the scanning, pulling, assembling, and routing that is tedious for a person precisely because it is routine. That is a large and valuable category. It is not the same as judgment. An agent can assemble the numbers and flag what looks unusual; deciding what the numbers mean and what to do about them is still yours.

The No-Code Builder

The point of the builder is in its name: you create and configure these agents without writing code. You describe what you want the agent to do — the task, the apps it should reach, the schedule, the output — and Vibe assembles the agent from that description, in plain language rather than a programming interface. This is deliberate and it matters, because the people who most need a recurring task automated — the operations manager, the analyst, the team lead drowning in routine — are usually not the people who can write a script to do it. The no-code builder puts agent-building in their hands.

In practice, building an agent is closer to delegating to a capable assistant than to programming. You state the job and its boundaries; you say which apps it may touch; you set when it runs; you describe what the output should look like and where it should go. You can then refine it the way you would refine instructions to a person — too noisy, tighten what it flags; missing something, tell it to include it. The skill is not coding. It is the older skill of describing a task clearly and setting its limits, which is exactly the skill this book keeps returning to.

Building One, Step by Step

Walk through building that Monday inbox agent from scratch, because the steps generalize to almost any agent and the order of them is part of the discipline. You begin with a **description in plain language**: "Every Monday at 8 a.m., look at the email that arrived over the weekend, summarize the main threads, and flag anything that looks time-sensitive — a deadline, a direct question to me, anything marked urgent. Post the summary to me; do not reply to anything." Notice how much of that sentence is *boundary* rather than task: "post the summary to me; do not reply to anything" is the clause that keeps the first version read-only, and writing it explicitly is better than assuming the agent will infer it.

Next you **grant the connectors it needs and no others**. The job touches email, so you connect email — and you connect *only* email, not your calendar or your drive, because an agent with access it does not need is a larger blast radius for no benefit. Then you **set the trigger**: a schedule, Monday at 8 a.m. Then you **describe the output**: a short summary, threads grouped, time-sensitive items called out at the top. Finally — and this is the step people skip — you **read the plan it shows you before the first run**. Vibe will lay out what it intends to do. The first time I built this, the plan revealed it was going to summarize *every* message including the newsletters and automated notifications, which would have buried the two messages that actually mattered. I added one line — "ignore automated notifications and bulk mail" — and the plan changed. That adjustment took thirty seconds and happened *before* the agent ever ran, which is exactly where you want to catch it.

The refinement loop after that is conversational. The first week's summary was slightly too long; I told it to be terser. The second week it flagged something as urgent that was not; I told it what "time-sensitive" actually meant to me with an example. Within a few iterations it had settled into something genuinely useful. The whole process never touched code, but it did demand the thing code cannot supply: a clear head about what the task actually is and where its edges are.

Realistic Capabilities and Limits

A guide that sold you on agents without the limits would be doing you a disservice, because the failure modes of an agent are more consequential than those of a chatbot — an agent acts, and it acts when you are not in the room.

An agent inherits every limit of the model underneath it. Everything true in the earlier chapters is true here. It can misread a document, misjudge what is urgent, summarize a source incorrectly, draft a reply that gets the tone or the facts wrong. The difference is that an agent may then *act* on that mistake — route the flawed summary, send the off-base draft — rather than showing it to you first. The model's fallibility does not vanish because you automated it; it gets a longer reach.

An agent does exactly what you configured, including the parts you got wrong. If your instructions are ambiguous, the agent resolves the ambiguity its own way, every time it runs, without checking. A human assistant who is unsure asks; an agent guesses and proceeds. A small misunderstanding in the setup becomes a recurring error you may not notice for weeks.

Connected actions touch real things. This is the heart of the risk and the reason the next section exists. An agent that only reads and summarizes is low-stakes — the worst case is a bad summary you can ignore. An agent that *sends, posts, files*, or *changes* something is operating on your real inbox, your real channels, your real records, and its mistakes land there. The blast radius of an agent is set by what you let it do, not by how careful the model is.

Automation runs unwatched. The whole value of an agent is that it runs without you. The whole risk of an agent is the same sentence. A mistake made while you are watching is caught in seconds; a mistake made by a scheduled job at 8 a.m. while you sleep is caught whenever you next look — which is why what an agent is allowed to do unattended should be scoped accordingly.

Consider how these limits compound rather than add. A plausible failure: you build a Friday agent that pulls support tickets and posts a summary to a team channel. The model misreads one ambiguous ticket (limit one), your instructions never specified how to handle tickets with no category so it guesses "general" every time (limit two), it posts directly to the channel rather than to you (limit three), and it does this every Friday while you are in your weekly review meeting and not looking (limit four). Any one of those alone is harmless. Together they produce a recurring, public, unwatched miscategorization that your team quietly starts treating as fact. None of the four is exotic; each is the ordinary behavior described above. The lesson is not that agents are dangerous in the abstract — it is that the failure modes stack, and the way you keep them from stacking is to break the chain at the cheapest link: keep the consequential action (posting publicly) behind your review until the agent has earned its way past it.

The Habit: Review the Plan

The discipline that makes agents safe to use is specific to agents, and it is this: **review the plan before the agent executes, and keep the consequential actions on a short leash.** Vibe is built to support exactly this — it plans the task and lets you review that plan before it acts, which is the agentic version of reading the code before you run it. Use that review; do not click past it.

Concretely, a few habits keep agents trustworthy. **Read the plan the first time, and after any change** — understand what the agent intends to do, which apps it will touch, and what it will send or change, before you let it run. **Start agents read-only.** Let a new agent scan, summarize, and report — work whose worst case is a bad summary — and live with it for a while before you ever let it send or change anything on its own. **Keep a human in the loop for actions with consequences.** An agent that *drafts* a reply for you to approve is a different and safer thing than one that *sends* the reply unseen; prefer draft-and-approve for anything that touches another person or a record that matters. **Scope the agent's reach to the task.** Give it access to the apps and data the job requires and no more, so a misunderstanding has a small radius rather than a large one. **Check the output for a while, even after it works.** An agent that has run cleanly for a month can still drift when the inputs change; the occasional spot-check is cheap insurance against a quiet, recurring mistake.

The frame to carry out of this chapter is the Monday-morning one. Agents are real leverage on the tedious, recurring, connective work that eats your time and asks nothing of your judgment — and handing that work to a machine you have scoped and supervised is a genuine win, not a gimmick. The leverage is proportional to the trust you extend, and the trust should be earned in stages: read-only first, draft-and-approve next, fully autonomous only for the low-stakes and well-proven. You stay the one who decides what the agent is allowed to do and who reads what it produces. The agent does the scanning and assembling; you keep the judgment and the responsibility — which, by now, you will recognize as the one rule this entire book is built around.

That rule is about to get its sharpest test in Vibe Code, where the machine writes software and the discipline of reading what it produces, before you trust it, becomes the difference between a faster engineer and a faster way to ship a bug.

Part IV: Vibe Code

Audience: developers and the technically comfortable. The deep parts are signposted.

"A coding agent that meets you in the terminal, the editor, and the cloud is one you actually reach for — but it never relieves you of reading the code it writes."

— A working principle

Chapter 12: Coding in Vibe

I asked Vibe Code to write a function that parsed a date format I always have to look up, the kind of fiddly string-handling that is not hard, exactly, just annoying enough that I put it off. It produced the function in seconds, and it looked right. I read it line by line — which took longer than reading the prompt I had typed — and found that it handled the common case cleanly but quietly mishandled a two-digit year in a way that would have been fine in testing and wrong in production six months later. I fixed the one branch, ran it against a fistful of edge cases, and shipped it. The generation saved me fifteen minutes of tedium. The reading saved me from a bug that would have cost a great deal more than fifteen minutes whenever it finally surfaced. That trade — fast generation, mandatory reading — is the entire chapter, and it is the heart of using Vibe Code well.

This is the first chapter of Part IV, on Vibe Code, the mode of Vibe built for software development. Vibe Code is a capable coding assistant, powered by Mistral's models — the Medium-class and coding-tuned variants behind Vibe, though the exact model names move quickly and you should confirm the current lineup against Mistral's documentation. It generates, debugs, and explains code, and it can run real work in cloud sandboxes rather than only talking about code in a chat box. It is not a magic wand, not a substitute for understanding your own code, and not a replacement for the engineer in the loop. This chapter is about what it does well, where its lane ends, and the one discipline you cannot skip.

What Vibe Code Does Well

Three things, mapping to most of what you actually want from an assistant while programming.

Generating code. Describe what you want — a function, a script, a class, a query, a configuration — and Vibe Code produces a working version, often a good one, usually faster than you would type it. It is strongest on the well-trodden: common languages, standard libraries, the routine plumbing that has to exist but does not need your cleverness. Boilerplate, scaffolding, the date-parsing chore — this is where generation saves the most, because it is exactly the work you would rather not do by hand.

Debugging. Give it code and an error, or describe the misbehavior, and Vibe Code is genuinely useful at finding the cause. The fresh-eyes effect is real: the model is not tired, has not stared at the file for an hour, and has seen thousands of versions of your bug. For the off-by-one, the missed edge case, the shadowed variable, the wrong type, the misused API — the everyday bugs — it is fast and frequently right, and "here is the error, what is wrong" is one of the highest-value things you can ask it.

Explaining. Underrated, and maybe the single most valuable coding use. Hand Vibe Code an unfamiliar block — a function in a codebase you inherited, a regex you cannot parse, an idiom in a language you are learning, an error message that makes no sense — and ask it to explain. It walks you through what the code does and why, which is leverage for *understanding* rather than just *producing*. In unfamiliar territory, the assistant that helps you understand is often worth more than the one that writes for you, because understanding is what lets you do the next thing yourself.

To see how these three uses interleave in a real session rather than as separate features, take a small but realistic task: you have inherited a script that reads a CSV of orders and is throwing an error on certain rows. You start with *explaining* — paste the function, ask "what does this do and why might it fail on some rows?" — and Vibe Code walks you through it and notices the code assumes every row has a value in the quantity column. That is a hypothesis, not a diagnosis, so you move to *debugging* — paste the actual error and a row that triggers it — and it confirms the empty-quantity case and points at the exact line. Then you move to *generating* — "add handling so empty quantities default to zero and log a warning" — and it produces the guarded version. Three uses, one continuous task, and notice that the explaining came first and did the heavy lifting: by the time you asked for the fix, you understood the bug well enough to know whether the fix was right. That ordering — understand, then diagnose, then generate — is a good default, because it keeps you ahead of the code the assistant is writing rather than chasing it.

The explaining use deserves one more word because it quietly compounds. Every time you ask Vibe Code to explain an idiom in a language you are still learning, you are not just unblocking the current task — you are building the knowledge that means you will not need to ask next time. Used this way the assistant is a tutor that happens to be sitting inside your work, and the developers who get the most out of it over months are the ones who treat explanation as the main event rather than a consolation prize when generation fails.

Cloud-Sandbox Remote Sessions

One thing distinguishes Vibe Code from a plain coding chat, and it is worth understanding because it changes what the tool can actually do. Vibe Code can run **remote sessions in cloud sandboxes** — isolated environments, hosted by Mistral, where code can be executed rather than merely written. Instead of handing you a block of code and trusting you to run it somewhere, Vibe Code can run it in a sandbox: execute the script, see the output, hit the error, iterate toward something that actually works.

This matters for two reasons. First, executed code is checked code in a way that generated-and-handed-over code is not — running it surfaces the errors that reading alone might miss, so the model can iterate against reality rather than against its own confidence. Second, the sandbox is isolated, which means this execution happens away from your machine and your systems, a meaningful safety property when you are running code you did not write. We will get into the practicalities of starting, saving, and moving between sessions — including the terminal and editor surfaces — in Chapter 13. For this chapter, the point is conceptual: Vibe Code does not just suggest code, it can run it in a controlled environment, and that makes it a more capable and a more careful collaborator. It does not, however, relieve you of reading what it ran. A sandbox tells you the code executed; it does not tell you the code is correct for your purpose.

The Limits, Stated Plainly

A guide that sold Vibe Code as a flawless programmer would be lying to you, and the honesty matters more here than almost anywhere, because broken code looks exactly like working code right up until it runs — and sometimes for a while after.

It can be confidently wrong. Vibe Code will produce code that looks correct, reads correct, and is subtly broken — an edge case unhandled, a condition inverted, an API used in a way that compiles but misbehaves. The fluency is the trap. Plausible-looking code carries no guarantee of correct code, and the model's confidence tells you nothing about whether it works. My date function looked perfect.

It can invent things. Asked to use a library, Vibe Code may call a function that does not exist, pass an argument the real API never took, or assume a behavior that is not the behavior. It is reconstructing a plausible API from training, not reading your installed version, and plausible is not real. This is the failure to watch for hardest, because it sails past a quick read.

It does not know your specifics. Vibe Code sees what you put in front of it — not your full codebase, your exact dependency versions, your conventions, your constraints, unless you supply them. Code correct in the abstract can be wrong for *your* project: the wrong pattern for your architecture, a version mismatch, a violation of a convention it had no way to know.

This last limit is the one you can do the most about, because it is largely a function of what you tell it. The cheapest improvement in the quality of what Vibe Code produces is feeding it the context it is otherwise guessing at. If your project uses a particular framework version, say so. If you have a house convention — errors returned rather than thrown, a specific logging call, snake_case over camelCase — state it, or better, paste a representative existing function so the model can match the style by example. The difference between "write me a function that uploads a file" and "write me a function that uploads a file, matching the error-handling and logging style of this existing handler I am pasting" is the difference between code you will spend ten minutes reshaping to fit and code that drops in. The model cannot read the room it is not shown. A surprising amount of what people experience as "the AI does not understand my project" is really "I asked it to work blind and it did."

There is a related failure worth naming because it is so easy to walk into: asking for something against a library version the model predates or postdates. The model reconstructs APIs from its training, and libraries change — a function gets renamed, an argument is deprecated, a default flips. If you are on a version the model has not internalized, the generated call can be confidently wrong in a way that looks completely ordinary. The defense is the same as for invented APIs: when the code calls into a dependency, check the call against the real documentation for *your* installed version, not against the model's recollection of how the library used to work.

Be honest about comparisons. Vibe Code is competitive with other coding assistants, and on Mistral's open and efficient models there are real reasons to choose it — cost, control, the sovereignty story this book keeps returning to. But I am not going to hand you invented benchmark numbers, because the figures move and a fabricated comparison is worse than none. Judge it on your own work, on tasks you can verify, rather than on a leaderboard claim from either side.

The Read-the-Code Discipline

This is the non-negotiable, the coding form of the rule that runs through every part of this book: **never run or ship code you have not read and understood.** Not "skimmed." Read, and understood well enough to know why it works.

The temptation is obvious and strong, and the sandbox sharpens it — the code looks right, it is right there, Vibe Code can even run it for you, and reading it carefully is work. But code you do not understand is a liability you have chosen to accept blind, and the failure modes above make that liability real, not theoretical. The cost of accepting broken code is not a polite error; it is a bug shipped into something that matters, a security hole, data corrupted, a problem that surfaces later when it is expensive — like a two-digit-year branch that works in every test and fails in production.

So the discipline, concretely. **Read every line before you trust it.** Understand what it does and why; the line you cannot follow is precisely the line to stop on — ask Vibe Code to explain it, or work it out yourself, but do not rely on what you cannot read. **Verify the unfamiliar APIs.** If it called a function or used a library you do not know cold, check the real documentation that it exists and behaves as claimed; this catches the invented-API failure directly. **Test it yourself**, beyond whatever the sandbox showed you — exercise the edge cases, confirm it does the thing on the inputs that matter. The model's confidence is not evidence; the sandbox's "it ran" is not the same as "it is right"; your test is the safety net, and it is the one part you cannot delegate.

What the reading actually catches is worth spelling out, because "read it" sounds like a posture rather than a procedure. When you read generated code with intent, you are hunting specifically for the failure modes above. You scan the boundary conditions — what happens on empty input, on the maximum value, on the negative case, on the off-by-one at the edge of the loop — because that is where confidently wrong code hides, looking fine in the middle of the range and breaking at the ends, exactly like my two-digit-year branch. You stop on any function call you do not recognize and ask whether it is real, because that is where invented APIs live. You check the code against your own project's shape — does it use the pattern you use, the version you have, the convention your team follows — because that is where abstractly correct code goes wrong for your specific situation. Reading is not a vibe; it is these three sweeps, and a developer who knows what to look for reads faster and catches more than one who is merely scanning for anything that "looks off."

The sandbox tempts you to substitute "it ran without error" for "I read it," and it is worth being precise about why that substitution fails. Running code proves it does not crash on the inputs you happened to run; it proves nothing about the inputs you did not. My date function ran cleanly against every test I would naturally have written, because the two-digit-year case is exactly the kind of input you do not think to try until it bites you in production. Execution and reading catch different things — execution finds the failures your test inputs trigger, reading finds the failures your test inputs miss — and you need both. The sandbox is a genuine asset precisely because it lets the model iterate against real execution, but it iterates against the cases it thought to run, which are not necessarily the cases that will hurt you. Your reading is what covers the gap.

This discipline is also how you stay an engineer rather than a person who pastes. If the work is to land in something real, you are responsible for it — and you cannot be responsible for code you do not understand. Vibe Code is a strong, useful companion that makes you faster at writing, fixing, and understanding code, provided you stay in the loop and read what it gives you. Take the speed and the second pair of eyes. Keep the judgment and the responsibility yourself, because the moment the code ships, it is yours.

That discipline does not change when Vibe Code leaves the chat box for the places developers actually work — the command line, the editor, the web. Only the surfaces do.

Chapter 13: The CLI and Editor Integrations

A coding assistant you have to context-switch into is one you reach for half as often as one that meets you where you already are. That is the whole argument for surfaces. Picture the difference: copying code out of a chat window and back into your editor, tab by tab, versus asking the question right there at the command line — in the directory you are already standing in, against the project you are already in the middle of, the answer coming back into the terminal where the work already lives. The friction that makes the chat-box version feel like a toy simply drops away. This chapter is not about what Vibe Code can do, which we covered, but *where* it lives.

This chapter walks through the surfaces Vibe Code offers — the command-line tool, the editor extension, and the web interface — and how work moves between them, including saved sessions and the cloud sandboxes from Chapter 12. Product specifics in this area move quickly, and one feature in particular is dated below as a near-future arrival; I flag those so you can confirm them against Mistral's current documentation rather than trust a printed date. The shapes, though, are stable enough to plan around.

The CLI Tool

The command-line tool is Vibe Code for people who live in the terminal, and for a great many developers that is most of the day. The CLI brings the assistant to the place where you already run your code, manage your files, and drive your version control, so asking it for help is not a context switch — it is a command, in the same window, against the project you are already in.

The reason this matters more than it sounds is friction. An assistant in a separate browser tab is one you use for the deliberate, sit-down question and forget about for the dozens of small ones that come up minute to minute. An assistant at your prompt is right there for all of them — the quick "what's wrong with this," the "write me a script for X," the "explain this error" — in the flow of the work rather than beside it. The CLI is also the natural home for the heavier, longer-running tasks, the ones that benefit from running in a cloud sandbox (Chapter 12) rather than blocking your own machine, which is where the next pieces come in.

A flavor of the register, kept deliberately minimal and illustrative — confirm the exact command names against current docs, as these move:

```
# Ask in the directory you're already working in
vibe "why is this test failing?"

# Kick off a longer task in a cloud sandbox
vibe run "add input validation to the upload handler"
```

The point of showing this is not the syntax, which will drift, but the feel: you stay at the prompt, the request goes from there, and the work happens either locally or in a sandbox without you leaving the terminal.

There is a deeper reason the CLI suits the work, beyond convenience: the terminal is where *context* already lives. When you ask a question at the prompt, you are standing in a directory full of the project's files, with its version-control state right there, in an environment that already knows your tools and paths. An assistant invoked in that setting can be pointed at the files around it without your having to copy them into a chat box — the context that the previous chapter warned the model lacks is, here, sitting one directory away. That is why the terminal-shaped tasks feel so natural through the CLI: the question "why is this test failing" is far more answerable when the test, the code it exercises, and the failure output are all reachable from where you asked, rather than pasted out of their habitat into a window that knows nothing else about your project.

The other thing the CLI is good for is the **long-running, hand-it-off task**. Some requests — "refactor this module," "add a test suite for this file," "trace this bug through the call stack" — are not quick questions; they take minutes and a lot of back-and-forth with the code. Running those in a cloud sandbox launched from the CLI means they do not tie up your machine and do not block the work you want to keep doing in the same terminal. You fire the task off, keep working, and come back to it. That pattern — quick questions answered inline, heavy tasks dispatched to the sandbox — is most of how a terminal-centric developer actually uses the tool day to day.

The VS Code Extension

Not everyone lives in the terminal, and for those who work primarily in an editor, the **VS Code extension** brings Vibe Code into that environment instead. The advantage here is context and proximity: the assistant sits beside your actual files, can work against the code you have open, and folds its suggestions and explanations into the editor you are already reading and writing in, rather than in a window off to the side.

This is the surface for the work that is naturally file-centric — generating code into a file, explaining a block you have selected, working through a change in the context of the surrounding code. Where the CLI suits the terminal-driven, command-shaped task, the extension suits the editor-driven, file-shaped one. Many developers will use both depending on what they are doing in a given hour, and the two are not rivals so much as the same assistant wearing the clothes of two different workplaces.

The editor's particular strength is **selection-scoped work**. Highlight a function and ask what it does; highlight a confusing block and ask it to be rewritten more clearly; highlight a failing chunk and ask what is wrong with *this*, specifically. The selection gives the assistant a precise target and gives you a precise place to read the answer — the explanation or the suggested change appears next to the very code it concerns, not in a separate conversation you have to mentally map back onto your file. That proximity is the whole point: the cost of checking the assistant's work, which Chapter 12 insists you must always pay, is lowest when the answer and the code sit side by side and your eyes barely have to move between them.

A worked feel for it: you are reading an inherited file and hit a dense, clever-looking block you do not trust. In the editor you select it, ask for an explanation, read the walkthrough in the panel beside it, decide the cleverness is hiding a bug, select the same block again, and ask for a clearer rewrite that you then read line by line before accepting. The entire loop — find, understand, judge, change, verify — happens without your leaving the file or copying anything anywhere. That tight loop is why developers who work mostly inside an editor tend to prefer the extension even when they also have the CLI installed: for file-shaped work, keeping the assistant in the file removes the friction that would otherwise make them reach for it less often.

The Web Interface

The third surface is the **web interface** — Vibe Code in the browser, requiring nothing installed. This is the most accessible entry point and the natural place to start: try the capability, run a self-contained task, work from a machine where you have not set up the tooling, or simply prefer a browser. It is the lowest-friction way in, and for plenty of tasks it is all you need. Where it gives ground to the CLI and the extension is integration — it does not sit in your terminal or your editor the way those do — but as a no-setup front door and a perfectly capable workspace in its own right, it earns its place among the three.

Choosing Among the Three

With three surfaces in hand, the natural question is which to use, and the honest answer is that the right choice follows the shape of the task more than any fixed preference. Reach for the **web interface** when there is nothing installed and you want there to be nothing installed — a quick trial of the capability, a self-contained task on a borrowed machine, the first time you ever touch the tool. Reach for the **CLI** when the work is command-shaped and lives in a project you are already standing in at the terminal: the inline questions, the version-control-adjacent tasks, the long jobs you want to dispatch to a sandbox and walk away from. Reach for the **editor extension** when the work is file-shaped and selection-scoped: explaining or rewriting a specific block, generating into a specific file, working a change in the context of the surrounding code where you can read the answer beside the thing it concerns.

Most developers settle into using two of the three heavily and the web interface occasionally, and the split usually tracks where they already spend their day. A terminal-first developer leans on the CLI and dips into the editor for the file-centric moments; an editor-first developer does the reverse. The mistake is to force one surface onto every task out of habit — copying code out of a chat tab when you could have asked at the prompt, or fighting the terminal for a change that wanted the editor's selection. Letting the task pick the surface is the small efficiency that, repeated dozens of times a day, is the whole reason the surfaces exist.

Switching Between Terminal and Cloud, and Saving Sessions

Here is the part that ties the surfaces together and is genuinely useful in practice: your work is not trapped wherever you started it. Vibe Code lets you **save sessions** and **move between the terminal and the cloud**, so a piece of work can begin in one place and continue in another without starting over.

Two ideas make this work. The first is the **cloud sandbox** from Chapter 12 — the isolated, hosted environment where code actually runs. Because the heavy execution happens in the cloud rather than on your laptop, a long-running task is not chained to your machine: you can start it and let it proceed in the sandbox. The second is **saved sessions** — the state of a piece of work persisted so you can come back to it. Together these mean you can kick off a task from the CLI in your terminal, have it run in a cloud sandbox, step away, and pick the session back up later — from the terminal again, or from the web interface — without losing the thread.

The practical value is real for the way work actually happens. Start something at your desk in the terminal, check on it from the browser on a different machine, hand a long sandbox job off and return to it after lunch. The session is the through-line, and the cloud is what frees the work from any single window. You are not re-explaining context or re-running from scratch each time you change surfaces; you are resuming.

A concrete day makes the value obvious. You arrive in the morning, open the terminal, and kick off a sandbox job to add a test suite to a gnarly module — the kind of task that runs for a while. You leave it running and go into a meeting. From your laptop in the conference room, you open the web interface, find the same session, and glance at how it is progressing without disturbing it. After lunch you are back at your desk and resume the session from the terminal again, read what it produced, and start working through the results. At no point did you copy state between windows, re-describe the task, or restart the work. The session carried the context across three surfaces and a hosted sandbox, and the only thing you brought along was your attention. That continuity is not a luxury feature; it is what makes the multi-surface story coherent rather than three disconnected tools that happen to share a name.

It is worth being honest about where this continuity has edges. The cloud sandbox is hosted by Mistral, which means a long-running session depends on that hosted environment being available and on whatever the sandbox can and cannot reach — it is isolated by design, which is a safety property, but isolation also means a sandboxed job does not automatically have your local secrets, your private network, or your machine's exact toolchain unless you have arranged for it. The sovereignty-minded reader should note the obvious corollary: code and context you push into a hosted sandbox are, for that moment, running on someone else's infrastructure, which is the same trade Chapter 1 laid out for the hosted assistant. For most coding work that trade is entirely comfortable. For the most sensitive codebases it is exactly the consideration that pushes a team toward the self-hosted path of Part V instead.

Slack-Started Jobs

One more surface is on the way, and I am dating it deliberately so you can check it: **starting Vibe Code jobs from Slack**, slated to arrive around **June 2026** as of this writing — confirm against Mistral's current documentation, because near-future features slip and the date is the most perishable thing in this chapter.

The idea is a natural extension of everything above. A lot of team coding work begins as a message — "can someone add validation to the upload handler," "we need a script for this." Slack-started jobs let that request kick off a Vibe Code task directly from the conversation where it was raised, running in a cloud sandbox the same way a CLI-launched job does, and bringing the work to where teams already coordinate rather than requiring someone to leave the conversation, open a tool, and re-type the ask. It is the same move the CLI made for the terminal — meet the work where it already is — applied to the place teams talk. Treat it as a capability to confirm rather than rely on until you have checked it is live, but the direction is clear and consistent with the rest of Vibe Code's design.

The Discipline Does Not Move

A closing word, because the surfaces are new but the rule is not. Everything in Chapter 12 still holds, on every surface in this chapter. Code that Vibe Code generates in the editor, runs in a sandbox from the CLI, or produces from a Slack message is still code you have not written, and the read-the-code discipline travels with it unchanged: read it, understand it, verify the unfamiliar APIs, test it yourself before you trust it. If anything, the smoother integrations raise the stakes, because the easier it becomes to go from "ask" to "running in my project," the more tempting it is to skip the reading in between. The CLI, the extension, the web, the sandbox, the coming Slack jobs — all of it makes you faster at getting to code. None of it makes you less responsible for the code you get. The surfaces are a convenience; the judgment stays yours.

That balance — real leverage held together with real responsibility — is the thread this book has been pulling since Chapter 1, and it carries straight into the next part, where we leave Vibe the assistant behind and turn to the open-weight models underneath it: what "open weight" actually means, and why, for the people this whole book was written for, it is the part that matters most.

Part V: Open Weights and Sovereignty

Audience: everyone who cares where their data goes — and the heart of why Mistral exists.

"An open weight is the difference between renting a capability and owning the ability to run it. For the people who can't rent, that difference is everything."

— A working principle

Chapter 14: What "Open Weight" Actually Means

A developer I know spent an afternoon arguing with a procurement officer over a single word. He wanted to deploy a Mistral model inside his company's own data center, and he had described it, in his proposal, as "open source." The procurement officer who had been burned before pulled up the license, read it line by line, and pointed out, correctly, that it was not the kind of open source she had approved software under in the past. There was no public training dataset. There was no recipe to rebuild the model from scratch. What there *was* turned out to be exactly what the developer needed: a file of trained parameters he was permitted to download, inspect, run, modify, and deploy on his own machines without sending anyone a byte of company data. The capability he wanted was entirely real. The label he had reached for was wrong. And the gap between the two is the single most important thing to understand about Mistral.

This chapter is about that gap. "Open weight" is the phrase that makes Mistral different from the American flagships, and it is also the phrase most likely to be misused, oversold, or quietly disappointing if you assume it means more than it does. I want to give you the honest version: what open weights genuinely let you do, what they pointedly do not guarantee, what the specific licenses permit, and why — once you strip away both the hype and the cynicism — this is still a meaningfully more open arrangement than anything the US labs offer, with real and shifting limits you should know about before you build on it.

Weights Are Not Source Code

Start with what a model actually is, because the whole confusion lives here.

When a lab trains a large language model, it ends up with two very different kinds of artifact. The first is everything that *went into* the training: the data the model learned from, the code that did the training, the configuration, the long and expensive process of turning raw text into a working model. The second is the *result* of all that — a very large file of numbers, the trained parameters, called the **weights**. The weights are the model. Load them into the right software and you have a working AI; they are the frozen distillation of everything the training taught it.

"Open source," in its traditional and demanding sense, means you get the *recipe* — enough of the source materials and process that someone else could, in principle, reproduce the result. "Open weight" means you get the *cake*. You receive the finished model — the weights — but not necessarily the training data, not necessarily the full training code, not the means to rebuild it from nothing. You can use it, study it, and change it. You generally cannot reconstruct how it came to be.

That distinction sounds academic until you notice how much it changes in practice. With the weights in hand you can do almost everything you would actually want to do with a model. What you cannot do is audit its origins the way a true open-source purist would demand, or independently verify exactly what it was trained on. For the developer in procurement, that trade was completely acceptable. For some buyers — a few regulators, a few security teams — the provenance question matters more, and "open weight" does not fully answer it. Knowing which kind of buyer you are keeps your expectations honest.

The cake metaphor is worth pressing a little further, because it explains a confusion people keep having. You can taste the cake, slice it any way you like, even add your own frosting on top — that is using, inspecting, and fine-tuning the model. What you cannot do is read the recipe and know precisely which flour was used, in what proportion, from which supplier — that is the training data and process. Now, for most purposes the recipe does not matter: you wanted a cake to serve at dinner, and you have one. But if you are a health inspector who needs to certify exactly what went into it, "here is the finished cake" is not the answer to your question, however good the cake is. This is why two reasonable buyers can look at the same open-weight release and reach opposite verdicts. The engineer who needs to *deploy* a capable model privately has everything; the auditor who needs to *certify provenance* has a real gap. Neither is wrong about the model. They are asking different questions, and "open weight" answers the first cleanly and the second only partially.

What Open Weights Genuinely Let You Do

Set the philosophy aside and look at what changes the moment you can hold the weights. This is the concrete, load-bearing list — the reason the whole arrangement matters.

You can download and run the model on your own hardware. This is the headline. The model lives on your machines, in your data center, behind your firewall. Nothing is transmitted to Mistral or anyone else when you use it. For the hospital lead from Chapter 1, for the bank, for the law firm, this is the difference between "forbidden" and "deployable." Chapter 15 covers the practical how.

You can inspect it. The weights are right there. Researchers can probe the model's behavior, security teams can test it in isolation, and you are not reasoning about a black box you reach only through someone else's API. You will not learn what it was trained on from the weights alone, but you can examine how it behaves under any condition you like, privately, as long as you want.

You can fine-tune it. Because you have the actual parameters, you can continue training the model on your own data to adapt it to your domain, your tone, your terminology, your task. The model becomes *yours* in a way a rented API model never can. Chapter 20 is devoted to this.

You can deploy it on your own terms, permanently. No usage caps you didn't set, no surprise deprecation of the version you depend on, no provider deciding to change the terms underneath you. You hold a copy. If Mistral changed its API pricing tomorrow, the model already running in your data center would keep running exactly as before. That durability — freedom from lock-in — is worth as much to some buyers as the privacy.

The common thread is **control**. Every closed, API-only model puts the provider between you and the capability. Open weights remove that intermediary entirely. You are not asking permission each time you use the model; you have the model.

It is worth seeing how these four capabilities combine for a real buyer, because in practice they are rarely wanted one at a time. Take a mid-size insurer evaluating an AI assistant for processing claims. The download-and-run property means the claims — which contain medical and financial details the insurer is legally bound to protect — never leave its network, which clears the regulatory bar that would have killed an API-only option outright. The inspect property means its security team can test the model's behavior in isolation and sign off, rather than reasoning about a vendor's black box. The fine-tune property means the insurer can adapt the model to its own claim categories and house language, so it performs on the actual work rather than on a generic benchmark. And the deploy-permanently property means the multi-year system it is about to build will not be undermined by a vendor deprecating the model underneath it. No single one of those four would have been enough; together they are the difference between a project that ships and one that dies in legal review. That bundling is the real shape of the open-weight advantage — not four separate features but one coherent answer to "can we actually build on this and keep control of our data."

What Open Weights Do Not Guarantee

Now the honest other half, because overselling this is how people get hurt.

Open weight does not mean fully reproducible open source. As above: you usually do not get the training data or the complete pipeline. If your definition of "open" requires the ability to rebuild the model from first principles, most open-weight releases — Mistral's included — do not clear that bar. They are open in the ways that matter for *use*, not necessarily in the ways that matter for *reproduction*.

Open weight does not mean uniformly open across every model. This is the part that trips people up most. Openness is not a single switch Mistral flips for its whole catalogue. Different models carry different licenses, and the terms genuinely vary. Some models are released under very permissive terms; others carry conditions; and at times Mistral has held a model back from open release entirely, offering it only through the API. The honest frame, the one from Chapter 1, holds here: *meaningfully more open than the American flagships, with real and shifting limits*. Do not assume that because one Mistral model is permissively licensed, the one you want is too. Check the specific model's license before you build on it.

Open weight does not mean unconditional. A permissive license is not the absence of a license. There are still terms — attribution, in some cases restrictions, in some cases carve-outs for certain kinds of use or certain kinds of company. "Permissive" means the terms are light and friendly to most users, not that there are none. The procurement officer in the opening was right to read the file.

What the Specific Licenses Permit

Two licenses come up most, and they sit at slightly different points on the openness spectrum. Both are perishable specifics — Mistral's licensing has shifted before and will shift again — so confirm the current terms on the model's own page before you commit. But as of this writing, here is the shape of it.

Mistral Large 3 — Apache 2.0. The flagship, the most capable model, carries one of the most permissive and widely trusted open licenses in software. Apache 2.0 is about as friendly as it gets: you can use the model commercially, modify it, redistribute it, and build on it, with essentially only standard attribution and patent-grant terms to honor. There is no revenue threshold, no special carve-out for large companies, no asterisk for production use. That the *most capable* model is also the *most openly licensed* is a genuinely striking choice, and it is the strongest single piece of evidence for Mistral's openness claim. It is also exactly the kind of thing that could change, so verify it.

Mistral Medium 3.5 — modified MIT. The efficient workhorse behind much of Vibe is released under a modified MIT license. Standard MIT is famously simple and permissive — do almost anything, just keep the notice. The *modified* part introduces exceptions, the notable one being conditions that apply to very-large-revenue companies, who may need a separate arrangement rather than relying on the open license alone. For the overwhelming majority of users — individuals, startups, mid-size firms, most enterprises — modified MIT behaves like the permissive license it mostly is. For a handful of the world's largest companies, the asterisk is real and worth reading carefully.

The practical takeaway: the two licenses are both permissive, but not identically so, and the difference is precisely the kind of detail "open weight" papers over if you let it. One model is unconditionally open; the other is open with a clause that matters to a small set of buyers. That nuance *is* the chapter.

How to Actually Read a Model's License

Because the licenses vary and the specifics move, the durable skill is not memorizing which model carries which terms — it is knowing what to look for when you open the license on a model's page, which you should do before building anything serious on it. A short checklist serves for almost any release. First, **commercial use**: does the license permit using the model in a product you sell, or is it limited to research and personal use? This is the clause that most often surprises people, because some "open" releases are open for tinkering but not for business. Second, **redistribution and modification**: may you pass the model on, and may you fine-tune it and distribute the result? For most uses you want both. Third, **conditions and carve-outs**: are there attribution requirements, use restrictions (certain industries or applications forbidden), or thresholds — like the large-revenue clause in modified MIT — that change the terms for certain companies? Fourth, **what travels with a fine-tune**: if you adapt the model, do the original license's terms bind the adapted version too? They usually do, and assuming otherwise is a common and expensive mistake.

The procurement officer in the opening did exactly this, line by line, and it is the reason the deal went through cleanly rather than blowing up after the fact. Reading the license is not lawyerly fussiness; it is the five-minute step that tells you whether the thing you are about to build your system on is actually yours to build on. And because the terms genuinely shift between model versions and over time, the read has to happen against the *current* license on the *specific* model you intend to use — not against your memory of a different model's terms, and not against what a blog post said last year.

Where You Get the Weights

Briefly, because Chapter 15 covers the practicalities: published model weights live in the places the machine-learning world already uses to share them — model hubs such as Hugging Face being the canonical example. You download the weights from there, point standard inference software at them, and run. The weights being *published* is what makes all of the above more than theory; it is the difference between a company saying it values openness and a company handing you the file.

Why This Matters Practically

It would be easy to treat this whole chapter as a licensing-lawyer's footnote. It is not. The open-weight question decides, concretely, who can use Mistral and how.

If you are an individual drafting emails in the Vibe app, "open weight" is mostly abstract — you are using a hosted assistant, and the licensing of the underlying model never touches you. But if you are the person deciding whether your regulated organization can adopt AI at all, the open-weight reality is the entire decision. It determines whether you can keep data in-house, whether you can avoid lock-in, whether you can satisfy a regulator who wants the model on infrastructure you control. A model you can download and run yourself clears bars that no API-only model can clear at any capability level, and a permissive license is what makes that legally clean rather than a risk your counsel will veto.

So hold both halves at once. Open weight is genuinely more open than the closed American flagships — download, inspect, run, fine-tune, deploy, no data leaving, no permission asked. And open weight is not the unconditional, fully reproducible, uniform openness the phrase can imply — licenses vary, provenance is partial, and the terms move. The buyers Mistral was built for do not need the maximalist version. They need exactly what open weight delivers: a capable model they are allowed to control, on terms light enough to actually use. The rest of Part V is about turning that permission into a running system — starting, in the next chapter, with the unglamorous question of what it actually takes to run one of these models on hardware you own.

Chapter 15: Running Mistral Models Yourself

A Mistral model running entirely on someone's own machine is a remarkably unremarkable thing to watch. There is no portal, no account, no API key, no terms-of-service checkbox. There is a folder of downloaded files on a server in the corner of the room, a few lines of command to start the inference software, and then a working AI answering questions — with the network cable, theoretically, unplugged. The whole sovereignty argument of this book, the entire reason a regulated hospital or a cautious bank can use Mistral when it cannot use the American flagships, comes down to that quiet, undramatic scene: a model, running on hardware the user owns, with nothing leaving the building.

This chapter is about making that scene real for you, or at least helping you decide whether you should. Chapter 14 established that you *can* obtain the weights and that the license permits you to run them. This chapter is the practical bridge from permission to a running system: how you actually get the weights, what hardware and software it takes, who genuinely benefits from doing this, and — just as importantly — who should not. The hands-on, command-level detail lives in Appendix C; here I want to give you the honest shape of the decision, because self-hosting is powerful and it is not free.

Getting the Weights

The mechanics of obtaining a model are, by design, mundane. Published open-weight models live in the model hubs the machine-learning world already uses to distribute them — Hugging Face being the canonical example — alongside their license, their documentation, and the various formats and sizes the model ships in. You go to the model's page, accept whatever license terms apply (this is where the Chapter 14 distinctions become concrete — read them), and download the weights.

What arrives is large. These are not small files; a serious model is many gigabytes, sometimes a great deal more, and you will often see several variants of the same model offered at different precisions and sizes. **Precision** here refers to how finely each parameter is stored — a full-precision model is bigger and slightly more capable; a *quantized* version compresses the weights to run on less hardware with a usually small quality cost. For many self-hosters, a quantized variant is the practical sweet spot, because it turns a model that needs a rack of GPUs into one that fits on far less. You do not need to master the theory to benefit; you need to know the option exists, because it changes the hardware math dramatically.

A worked picture makes the precision choice concrete. Suppose a model has a given number of parameters. Stored at full sixteen-bit precision, each parameter takes two bytes, so the raw weights alone occupy roughly two gigabytes per billion parameters — before you account for the extra memory the running model needs to actually hold a conversation. Quantize those same weights down to eight bits and you roughly halve that; down to four bits and you roughly halve it again. The capability does not fall in proportion: a well-made four-bit quantization of a strong model typically loses only a little quality, often little enough that a careful evaluation on your own tasks cannot reliably tell the difference, while the memory footprint shrinks by a factor that can decide whether the model fits on hardware you already own. This is why so much practical self-hosting advice comes down to "try the quantized version first." The instinct to insist on full precision because it is technically the best is, for most real workloads, the same mistake as insisting on the flagship when the mid-tier would do — you are paying for a margin you cannot measure.

One honest caveat travels with this: quantization quality is not uniform across methods or across models, and the only reliable way to know whether a given quantized variant is good enough is to run your real tasks through it and look at the output, not to trust a single benchmark number on the model's page. Treat the quantized version as the default to *evaluate*, not the default to *assume*. Run the same handful of representative prompts through the full-precision and the quantized variant side by side, read both answers, and let your own task — not a leaderboard — decide whether the smaller footprint costs you anything you care about.

The Hardware Reality

Here is the single most important thing to internalize, and the place where wishful thinking causes the most grief: **the hardware you need scales directly with the size of the model, and the range is enormous.**

At one end sits the flagship. Mistral Large 3 is a very large model — on the order of hundreds of billions of parameters — and running it well is a serious infrastructure undertaking. You are looking at multiple high-end GPUs, substantial memory, and the kind of setup that lives in a data center rather than under a desk. This is real money and real operational weight. It is entirely doable — organizations do it every day — but nobody should imagine they will casually spin up the full flagship on a spare workstation. The largest models demand commensurate hardware, full stop.

At the other end, Mistral's smaller models are genuinely modest in their appetite. The efficient members of the family are part of why Mistral matters: a smaller model, especially quantized, can run on a single capable GPU, and the most compact ones can run on hardware a serious enthusiast or a small team already owns. The capability of these smaller models has climbed steadily, so "small enough to run cheaply" no longer means "too weak to be useful" — for a great many real tasks, a well-chosen smaller model self-hosted is exactly the right answer.

Between those two ends sits the middle, and it is where most successful self-hosting projects actually live. A mid-sized model at a sensible quantization will often run on a single high-end GPU of the kind a serious team can buy or rent without a capital-expenditure committee, and serve a real workload — internal support triage, document classification, drafting — at a throughput that comfortably keeps up with the volume that workload produces. The art is not in heroics at either extreme; it is in finding the smallest model and the most modest hardware that clears your quality bar, because every step up in model size is a step up in GPU cost, power draw, and operational fragility.

It is worth being concrete about what "the hardware math" actually involves, because the parameter count is only the first term. You need memory not just for the weights but for the working context — the running state the model builds up as a conversation or document grows — and that working memory rises with how much text you feed the model at once and how many requests you serve in parallel. A model whose weights fit comfortably on a card can still run out of memory if you ask it to process very long documents or to handle many simultaneous users, because the context state grows on top of the fixed weight footprint. This is the most common way a self-hosting plan that looked fine on paper falls over in practice: the weights fit, the team declares victory, and then the first real workload — long inputs, concurrent users — pushes past the memory budget nobody sized for. Plan for the weights *plus* the working memory *plus* headroom, not the weights alone.

The whole spectrum is the point. **Match the model to the hardware you have, or the hardware to the model you need** — and resist the instinct to reach for the flagship when a smaller model would clear your bar at a fraction of the cost and operational burden. Chapter 19, on choosing a model, is the companion to this decision; the sizing question and the capability question are really the same question viewed from two sides.

The Software That Runs It

You do not run a model by double-clicking the weights. Between the file and a working answer sits an **inference framework** — software whose job is to load the weights, manage the GPU, and turn your prompts into the model's responses efficiently.

The good news is that this is a mature, well-trodden part of the ecosystem. Standard, widely used inference servers — vLLM is the most prominent example for production-grade serving, and there are others tuned for different scales and use cases — are built precisely to take a folder of open weights and serve them as a fast, concurrent API on your own infrastructure. They handle the hard parts: batching many requests together, using GPU memory well, exposing the model through an interface your applications can call. From your software's point of view, a self-hosted model behind one of these servers looks much like calling a hosted API — except the endpoint is a machine you own, and the data never leaves it.

It is worth distinguishing the two broad reasons people run a model locally, because they pull toward different tools. One is **convenience and experimentation** — a developer wants the model on their own laptop to try it, to keep a sensitive draft off any network, to work on a plane. For that, the friendly desktop runners (the kind that wrap everything behind a single install and a simple command, with quantized models a click away) are the right starting point; they are not built for serving an organization, but they are unbeatable for "I just want this running in front of me in ten minutes." The other reason is **production serving** — a model that has to answer many requests, reliably, behind an application that real users depend on. That is the territory of the production inference servers, which trade the one-click simplicity for the throughput, concurrency, and operational controls a real service needs. Confusing the two is a common early misstep: people try to build a production service on a desktop runner and find it does not scale, or they wrestle with a production server when all they wanted was to kick the tires. Pick the tool that matches the reason you are running the model, and the path gets much smoother.

Whichever camp you are in, the request you eventually send looks reassuringly ordinary. A production inference server exposes the model through a conventional, chat-style API — frequently one shaped to match the interface developers already know from the hosted providers — so the code that talks to your self-hosted model is, in most cases, nearly identical to the code that would talk to a hosted one. The endpoint changes from a vendor's URL to a machine on your own network; the secret key may vanish entirely because you control the box; but the request and the response keep the familiar shape. That portability is quietly important: it means you can prototype against a hosted API for convenience and later move the same application onto self-hosted weights with far less rework than you might fear, which is exactly the door Chapter 19's "openness" dimension keeps open.

Which framework, which exact configuration, and the command-line specifics are exactly what Appendix C is for. The durable point for this chapter is that you are not building this from scratch. The tooling to run open weights yourself is real, actively maintained, and used in production by serious organizations. The model being open is what makes it *possible*; this software ecosystem is what makes it *practical.*

Who Should Self-Host

Self-hosting is the right choice for a recognizable set of people and organizations. If you see yourself clearly in this list, the operational burden is almost certainly worth it.

Organizations that cannot let data leave. This is the core case the whole book keeps returning to. Hospitals, banks, law firms, government bodies, defense-adjacent work, anyone under data-residency rules they cannot negotiate — for them self-hosting is not an optimization, it is the only door. No hosted API, however well-secured, satisfies a requirement that the data physically never goes to a third party. Self-hosting does.

Teams that want to escape lock-in and control cost at scale. If you are running a very high volume of inference, the economics can flip: a model you host on hardware you own (or rent as raw compute) can be cheaper at scale than per-token API pricing, and it insulates you from a provider changing terms or deprecating the version you depend on. You hold the model. It keeps working.

Builders who need to customize deeply. If you are fine-tuning heavily (Chapter 20), embedding the model in an unusual pipeline, or doing research that requires probing the model directly, you want the weights in your hands and the model on your own infrastructure. The freedom to modify is the whole appeal, and it only exists when you self-host.

To make these abstractions less abstract, consider how the decision actually plays out for three recognizable self-hosters. A regional hospital that wants to summarize and route clinical correspondence is squarely in the first group: the data may not leave its network, so the only question is which model it can run on the hardware it can justify — almost certainly a mid-sized or small model, quantized, on a modest GPU server inside its own walls, because a model it is *allowed* to run beats a more capable one it is forbidden to touch. A consumer app with millions of daily classification calls is in the second group for a different reason: residency is not its problem, but per-token API pricing multiplied by that volume eventually crosses the line where owning the inference is cheaper than renting it, and self-hosting becomes a margin decision rather than a compliance one. And a research team probing how a model reasons is in the third group: it needs the weights in hand not to save money or satisfy a regulator but because the work is impossible through a hosted black box. Notice that all three reach the same place — a model on their own infrastructure — for entirely different reasons, and that the reason shapes everything downstream, from which model they pick to how much they are willing to spend keeping it running.

Who Should Not

Equally important, and less often said honestly: self-hosting is the *wrong* choice for a great many people, and choosing it anyway buys you cost and complexity for benefits you will not use.

If your data can use a hosted service, a hosted service is usually easier and often cheaper. Standing up and operating GPU infrastructure — capacity planning, scaling, uptime, security patching, the on-call burden when it breaks at

2 a.m. — is real work with real cost. If you do not have a residency requirement that forbids the hosted path, Mistral's own API (Chapter 18), including its EU-based infrastructure (Chapter 16), gives you the model without the operational weight. For individuals and most teams, that is simply the better trade.

If you do not have the hardware or the people, do not pretend the flagship is free because the weights are. "Open weight" means no license fee and no data leaving; it does not mean no infrastructure cost. The weights are free; the GPUs, the electricity, and the engineer who keeps it all running are not. Counting only the first and ignoring the rest is the most common way self-hosting projects disappoint.

It is worth dwelling on the failure modes here, because they are predictable and they are the ones that turn an enthusiastic pilot into a cautionary tale. The first is the **idle-GPU trap**: a team buys or rents enough hardware to run the flagship, then discovers their actual traffic uses a fraction of that capacity, so they are paying around the clock for compute that sits idle most of the day. A hosted API charges you only for what you use; a GPU you own charges you whether it is busy or not, and at low utilization the per-request economics of self-hosting can be far worse than the API the team was trying to escape. The second is the **operational-burden surprise**: the model runs fine in the demo, ships, and then the team meets everything a hosted provider was quietly handling — the security patches, the driver updates, the capacity planning, the monitoring, the page at two in the morning when the inference server falls over and the application that depends on it goes dark. The third is the **stale-model drift**: a self-hosted deployment does not update itself, so a team that stood up a model a year ago and never touched it again is now running something the rest of the field has moved past, with no automatic upgrade path and a migration they have been putting off.

None of these is a reason not to self-host; all three are reasons to go in with eyes open. The clean test is to ask, before committing, who owns each of those burdens on your team and whether that person has the time and skill to carry it. If the honest answer is "nobody, really," that is not a gap to paper over with optimism — it is the signal that the hosted path (Chapter 18), including Mistral's EU-based infrastructure (Chapter 16), is the better trade for you, and choosing it is not a failure of nerve but a correct reading of your own situation.

The Trade at the Heart of It

Strip everything down and self-hosting is a single trade, stated plainly: **maximum control and maximum data privacy, paid for with operational burden and infrastructure cost.**

On one side of the ledger you get the things nothing else delivers — the model on your own hardware, your data never leaving, no dependence on a provider's terms or uptime, the freedom to inspect and modify, durability against the model being changed or retired underneath you. On the other side you take on everything a hosted provider was quietly doing for you: the hardware, the scaling, the reliability, the security, the expertise. There is no version of self-hosting where you get the first column without paying the second. The question is never "is self-hosting good" in the abstract; it is "is the control worth the burden *for me*," and the honest answer depends entirely on whether you have a requirement that makes the control non-negotiable.

For the organizations Mistral was built to serve, the answer is yes without hesitation, because for them the control column is not a luxury but a legal prerequisite — and that is precisely the subject of the next chapter, where data residency and sovereignty move from the *how* of running a model to the *why* it changes who is allowed to use AI at all. For everyone else, this chapter is permission to choose the easier path with a clear conscience: the weights being open means self-hosting is *available* to you, not that it is *obligatory*. The whole spectrum exists so you can pick the point that fits — and knowing you could run it yourself, even if you choose not to, is itself a kind of freedom the closed flagships never offer.

Chapter 16: Data Residency and Sovereignty

A compliance officer at a European insurer once described her job to me as "knowing where every byte sleeps at night." She did not mean it poetically. Under the rules she answered to, it genuinely mattered which country a customer's data sat in while it was being processed, which company's hands it passed through, and whose legal system could compel access to it. When her colleagues got excited about an AI assistant that could automate half her department's drudgery, her first and only question was not "how good is it." It was "where does the data go, and who can reach it once it is there." For most of the impressive tools, the honest answer was: to servers in another country, in the hands of a foreign company, reachable in principle by a foreign government — and that answer ended the conversation before it began.

This chapter is about the answer Mistral can give that the American flagships structurally cannot. Sovereignty is the word the company uses, and it is easy to wave past it as marketing. It is not marketing. It is a precise, practical set of guarantees about *where your data lives, who can access it, and which laws govern it* — and for a large and important class of users, those guarantees are the entire difference between an AI they can adopt and one they are forbidden to touch. This is, I will argue, Mistral's defining advantage, the thing no benchmark captures and no US-based competitor can fully match.

What Sovereignty Actually Means

The word gets used loosely, so let me pin it to three concrete questions, because sovereignty is really the answer to all three at once.

Where does the data physically live? When you send a document to an AI to be summarized, that document is processed on a computer that sits in some physical place, in some country, under some jurisdiction. Data residency is the guarantee about *which* place. For a great many regulated organizations, the law is explicit that certain data must remain within a particular region — the EU, most commonly — and may not be exported. "Where the data lives" is not a preference; it is a line drawn by a regulator.

Who can access it? Beyond where the data sits, there is the question of whose hands can reach it. A foreign company processing your data is subject to its own government's laws — including, in some jurisdictions, laws that can compel that company to hand over data regardless of where the servers physically are. This is the quiet part that catches organizations off guard: data residency alone is not enough if the *company* holding it can be legally reached from outside your region. Sovereignty addresses both the location and the reach.

Which laws apply? Every arrangement sits inside a legal framework, and frameworks conflict. A US-based provider operates under US law; a European organization operates under EU law; and the two do not always agree about privacy, access, or government compulsion. Sovereignty means being able to keep your AI processing inside a single, predictable legal regime — yours — rather than straddling two that may pull in opposite directions.

Put together, sovereignty is the ability to answer the compliance officer's question cleanly: the data lives *here*, only *these* parties can reach it, and *this* law governs it — and all three answers are ones your regulator will accept.

The reason all three questions have to be answered at once, rather than one at a time, is that they fail independently. A provider can give you a flawless answer to the first — yes, your data is processed in a data center inside the EU — and still fail the second, because the company operating that data center answers to a government outside the EU that can, under its own law, compel the company to produce data wherever it physically sits. Location and reach are different guarantees, and a deployment that satisfies one while quietly failing the other is exactly the kind of arrangement that passes a casual glance and fails a serious legal review. The compliance officer's instinct to ask all three at once is not bureaucratic thoroughness for its own sake; it is the hard-won knowledge that a clean answer to "where" is worthless if the answer to "who can reach it" is "a foreign authority, on demand."

It helps to walk a single byte through the questions to see how concrete they are. A customer's insurance claim — a document full of medical detail and personal identifiers — is dropped into an AI to be summarized. *Where does it live?* For the few seconds it is being processed, it sits in the memory of a specific machine in a specific building in a specific country; if that country is outside the region the insurer's regulator permits, the answer fails before any other consideration. *Who can reach it?* The company running that machine can read it, and so can anyone that company is legally obliged to obey; if that includes an authority the insurer's own legal system does not recognize as having jurisdiction over its customers' data, the answer fails here instead. *Which law applies?* If the processing straddles two legal regimes that disagree about access and compulsion, the insurer is exposed to whichever one reaches furthest, not the one it chose. Only when all three answers land inside the insurer's own regime is the claim safe to process — and that, precisely, is what sovereignty is for.

Why US-Based Providers Cannot Fully Match This

This is the structural part, and it is worth being precise rather than tribal about it. The American AI flagships are excellent tools run by serious companies, and many of them now offer European data centers and respectable privacy terms. The issue is not competence or good faith. The issue is *structure*.

A US-headquartered company is subject to US law no matter where its servers sit. Even with European data centers and strong contractual promises, the company itself remains reachable through its home jurisdiction in ways that some European regulators and some risk-averse legal teams will not accept for their most sensitive data. The provider can promise a great deal, and mean all of it — but it cannot promise to be outside the legal reach of its own government, because no company can. For the insurer's compliance officer, that residual reach is exactly the part that fails her test.

Mistral's advantage is that it sits inside the European legal framework natively. It is a European company, offering European infrastructure, operating under European law. When it tells a European organization that data stays in-region under EU jurisdiction, there is no foreign legal system quietly reaching over the top. This is not Mistral being cleverer than the Americans; it is Mistral being *located where its customers need their AI to be located*. The advantage is geographic and jurisdictional, and it is therefore one a US-based competitor cannot replicate by building a data center, because the company's own nationality is the thing in question.

This is also where a great deal of well-meaning effort gets spent on the wrong remedy. Faced with a residency concern, the natural reflex is to reach for a contract — a data-processing agreement, a strongly worded clause, a promise of EU-only handling — and contracts genuinely matter. But a contract is a promise made *within* a legal system, and it cannot lift the company that signs it out of its own government's reach. A US-based provider can promise, in good faith and in writing, that European data stays in Europe and is never disclosed, and still be placed by its own government in the position of having to choose between honoring that promise and obeying a lawful order at home. The contract does not resolve that conflict; it only documents it. What the risk-averse legal teams are reacting to is not a suspicion that the provider is lying — it is the recognition that the provider is structurally unable to guarantee the one thing they need, no matter how sincerely it promises. Mistral's answer to that concern is not a better clause. It is the absence of the conflict in the first place, because there is no second jurisdiction reaching in.

A useful way to feel the difference is to imagine the same demand arriving at two providers. A lawful order lands, from an authority outside Europe, asking for data about a European customer. The US-based provider, however European its data center and however firm its contract, now has a genuine conflict to navigate between two legal systems, and the European customer's protection depends on how that conflict resolves — a resolution the customer does not control and cannot fully predict. The same order arriving at a European provider operating entirely under European law simply has no purchase: the authority is asking a company that does not answer to it, about data that never left a jurisdiction it does not govern. The point is not that one company is more trustworthy than the other. It is that one is exposed to a conflict the other structurally does not have, and for the most sensitive data, removing the conflict beats managing it.

The Two Ways Mistral Delivers It

Sovereignty with Mistral is not a single switch; it is a spectrum, and you choose the point that fits your requirement and your appetite for operational work.

EU-based hosted infrastructure. The lighter-weight option: use Mistral's models through its own European infrastructure, where the data is processed in-region, under EU jurisdiction, by a European company. You do not run anything yourself — you call a hosted service, much as you would any cloud AI — but the residency and jurisdiction questions come back with the answers a European regulator wants. For organizations whose requirement is "EU data residency under EU law" rather than "data must never touch a third party at all," this is often the perfect fit: most of the convenience of a hosted assistant, with the sovereignty guarantees built in.

Self-hosting, for the strictest cases. When the requirement is absolute — data must never leave our walls, full stop — the open weights make the strongest answer possible. You download the model and run it on your own infrastructure (Chapter 15), and there is no residency question at all because there is no transmission at all. The data is processed inside your own network, on your own hardware, governed by nothing but your own rules. This is the maximal end of the spectrum: maximum control, in exchange for the operational burden Chapter 15 laid out. For defense work, for the most sensitive health and financial data, for anyone who cannot accept *any* third-party processing, this is the only arrangement that clears the bar — and the open-weight model is what makes it available.

The same model family serves both ends. You are not choosing a different product for hosted-EU versus fully self-hosted; you are choosing where on the control spectrum you sit, and Mistral is built to let you sit anywhere on it. That flexibility is itself part of the sovereignty story.

The choice between the two is not a matter of which is more "secure" in the abstract — it is a matter of matching the arrangement to what your rules actually forbid, and the two rules look deceptively similar until you read them closely. "Our data must be processed in the EU under EU law" and "our data must never be sent to a third party at all" sound like neighbors, but they point to different ends of the spectrum. The first is satisfied by the hosted EU option: the data does travel to Mistral, but it travels to a European company processing it in-region under European law, which is exactly what that rule demands. The second is satisfied only by self-hosting, because *any* transmission to *any* third party, however European and however well-secured, violates a rule written as an absolute. The expensive error is to read the second rule, panic, and assume you must build a data center — when a careful reading of your own obligation shows the first, lighter option clears it. Read the rule before you choose the architecture; the architecture should follow the requirement, not the fear.

In practice the dividing line tends to fall along the sensitivity of the data rather than the size of the organization. The same insurer might run routine internal workflows — drafting standard correspondence, summarizing public filings — on the hosted EU service, while reserving the on-premises, nothing-leaves arrangement for the genuinely sensitive claims data, because the residency rule that governs the second is stricter than the one that governs the first. That a single organization can place different workloads at different points on the spectrum, according to what each workload's rules require, is not a complication to be tidied away; it is the whole advantage. A provider that offers only its own cloud forces every workload into one shape. Mistral lets the architecture follow the data.

Who Genuinely Needs This

It would be dishonest to tell every reader that sovereignty is their deciding factor. For many it is not, and I would rather you know which group you are in than absorb a fear that does not apply to you.

If you are an individual using Vibe to draft emails and brainstorm, sovereignty is mostly abstract. You are using a hosted assistant like any other; the privacy terms matter, and they are good, but you are not the person for whom residency is a legal line. Use the tool and enjoy it.

The people for whom this chapter is *the* chapter are recognizable. **Regulated industries** — healthcare, finance, insurance, law — where the handling of sensitive data is dictated by regulators rather than chosen by preference. **The public sector** — governments and public bodies that, increasingly, are expected or required to keep citizen data within national or European control. **Organizations under GDPR and the tightening EU regulatory environment**, where the obligations around data are escalating rather than relaxing — the EU AI Act's key provisions begin landing through 2026, and the broader direction of travel is toward more residency and accountability, not less (Chapter 23 owns the compliance detail; I will not over-specify it here, because it is exactly the kind of fast-moving thing that dates a book). And **any organization, anywhere, that has decided its most sensitive data simply will not be sent to a foreign black box** — a decision more and more boards are making on their own, regulation or not.

For all of these, the calculation is not "Mistral versus the American flagships on capability." It is "which capable AI am I actually *permitted* to use" — and once that is the question, the field narrows sharply, often to Mistral alone.

Why This Is the Defining Advantage

I want to end where Chapter 1 began, because this chapter is the proof of that opening claim. The American flagships ask you to choose between capability and control. For the compliance officer, for the insurer, for the public body, that is not an abstract trade-off — it is a wall, and on the wrong side of it sits every brilliant tool they are not allowed to use.

Sovereignty is the mechanism that gets those people over that wall. Because Mistral can keep the data in-region under European law, or let you run the model with nothing leaving at all, the capability becomes *reachable* for people who were locked out. And that reachability is not a feature you can put on a leaderboard or demonstrate in a flashy demo. It is quieter and more decisive than that: it is the difference between an AI strategy your compliance officer can sign and one she has to veto. A model your regulator will let you run beats a more impressive one she has to forbid — not as a slogan, but as the daily arithmetic of the people who live inside those rules.

That is why sovereignty, not raw benchmark capability, is the heart of Mistral's case. What remains is to turn that principle into a product an organization can actually buy and deploy — the private cloud, the on-premises options, the enterprise controls and support that wrap the sovereignty proposition into something a business can adopt with confidence rather than assemble by hand.

Chapter 17: Vibe Enterprise: Self-Hosting and Private Cloud

When a mid-sized European manufacturer finally decided to roll out AI across its workforce, the project did not stall on which model to use or what it could do. Everyone agreed the capability was there. It stalled, for the better part of a quarter, on a stack of questions that had nothing to do with intelligence. Who in our company can administer this? How do employees sign in — can it use our existing identity system, or do we create yet another set of logins? Which teams can see which data? Where does it run, given what our legal department said about residency? And if something breaks, who do we call? Those questions are not glamorous, and no demo ever answers them. But they are the questions that decide whether AI actually arrives inside an organization or remains a thing individuals quietly experiment with on their own accounts. This chapter is about the part of Mistral built to answer them.

Everything in Part V so far has been about *capability and principle* — what open weights permit, what running a model yourself takes, why sovereignty changes who is allowed to use AI. **Vibe Enterprise** is where those principles become a product an organization can actually buy and deploy: the assistant from Parts III and IV, plus the deployment options, administrative controls, security, and support that turn "you could self-host this" into "your company runs this, properly, at scale." It is the full sovereignty proposition packaged for institutions rather than assembled by hand.

The Deployment Spectrum

The defining feature of Mistral at the enterprise level is that you are not handed one deployment model and told to live with it. You are offered a *spectrum*, and the right answer depends on how much control your situation demands and how much operational work you are prepared to take on. It helps to picture three broad points along that line.

Hosted, on Mistral's infrastructure. The lightest-touch option: your organization uses Vibe and the underlying models through Mistral's own service, including its EU-based infrastructure for the residency guarantees of Chapter 16.

Mistral runs everything; you administer your users and your data, but the machines and their upkeep are not your problem. For organizations whose requirement is "EU data residency under EU law, with enterprise controls" rather than "nothing may ever touch a third party," this is frequently the sweet spot — most of the sovereignty, none of the infrastructure burden.

Private cloud. A middle path that has become the default for a great many serious enterprises. Here Vibe and the models are deployed into a cloud environment dedicated to your organization — your own isolated tenancy, your own boundary — rather than a shared service. You get strong isolation and a deployment that answers to your controls and your residency requirements, while still leaning on cloud infrastructure rather than building a data center. It is the balance point of the spectrum: substantially more control than the shared hosted service, substantially less operational weight than running everything on your own metal.

On-premises and fully self-hosted. The maximal-control end, building on Chapter 15. The models run inside your own data center, on hardware you own, behind your own firewall, with — for the strictest cases — nothing leaving your walls at all. This is the arrangement for the organizations whose requirements are absolute: defense, the most sensitive health and financial data, anyone who cannot accept third-party processing in any form. It is the most demanding to run, and for the buyers who need it, it is the only acceptable answer. The open weights are what make it possible; Vibe Enterprise is what makes it supported rather than a do-it-yourself project.

The distinction between the private-cloud middle and the on-premises end deserves more than a label, because it is where a lot of enterprise confusion lives. Both give you isolation; what differs is *who owns the metal.* In private cloud, the hardware still belongs to a cloud provider — you have your own dedicated, isolated tenancy on it, but you are renting infrastructure rather than buying it, which means you inherit the cloud's elasticity (scale up for a busy quarter, scale back down) without ever touching a physical machine. On-premises, the hardware is yours: bought, racked, powered, and maintained inside a building you control, which is the only arrangement that can satisfy a rule saying data must never reach any outside infrastructure at all. The practical consequence is that private cloud is often the right answer for an organization whose rule is "isolated and EU-resident" but whose appetite for running a data center is zero, while on-premises is reserved for the narrower set whose rule is genuinely "nothing leaves." Many organizations talk themselves into on-premises out of an instinct for maximum control and then discover, a quarter into the project, that private cloud would have cleared their actual requirement at a fraction of the operational weight.

The point of the spectrum is that **you choose where you sit based on your control requirements, not the vendor's convenience.** Most AI providers give an enterprise exactly one shape: their cloud, their terms. Mistral lets a single organization place different workloads at different points — a sensitive workload on-premises, a routine one in private cloud — and that flexibility is itself a large part of why institutions choose it.

What "Enterprise" Adds Beyond Deployment

Deployment location is only half of what makes a tool adoptable across an organization. The other half is the unglamorous machinery of running software for hundreds or thousands of people, and this is where Vibe Enterprise earns the name.

Administration and user management. Someone has to provision users, group them by team, set who can do what, and manage all of it as people join and leave. Enterprise deployment provides the admin layer to do that centrally, rather than leaving each employee to fend for themselves. Without this, an "AI rollout" is just a lot of individual accounts; with it, it is a managed system.

Single sign-on and identity integration. The question the manufacturer got stuck on. Enterprises do not want a separate set of credentials for every tool — they want employees to sign in through the identity system they already run, with the access policies that already exist there. SSO integration means Vibe slots into that existing identity fabric rather than standing outside it, which is both a security requirement and the difference between adoption and friction.

It is easy to underrate how much weight this single feature carries, so it is worth being concrete about what it actually buys. When Vibe authenticates through the company's existing identity provider, three things happen that are nearly impossible to achieve with standalone accounts. The first is that *deprovisioning works*: when an employee leaves and their central account is disabled, their access to Vibe disappears with it, automatically, rather than lingering as a forgotten login that a security review will eventually flag. The second is that *access policy is inherited*: the rules the organization already enforces — who must use multi-factor authentication, which groups exist, what a given role is allowed to touch — apply to Vibe without being rebuilt by hand, which is both less work and less opportunity for the two systems to drift apart. The third is that *the audit story is unified*: sign-ins to Vibe show up in the same place as every other access event, so a compliance team can answer "who used this, and when" from the systems they already trust. Without SSO, an AI rollout quietly creates a parallel population of accounts that nobody is governing, and that shadow population is exactly the kind of thing that turns a security audit from routine into painful. The manufacturer was not being fussy when it got stuck here; it was correctly refusing to bolt an ungoverned identity system onto the side of its carefully governed one.

Security and access controls. Beyond who can log in, there is the question of who can reach which data and what the audit trail looks like. Enterprise controls govern access at a granularity organizations need, so that the sensitive data stays visible only to the people entitled to see it, and so that security and compliance teams can verify that it does. For the regulated buyers of Chapter 16, these controls are not a nicety; they are part of what makes the deployment defensible to an auditor.

Enterprise support. And finally, the answer to "who do we call." Self-hosting open weights by yourself means you own every problem that arises. Vibe Enterprise wraps the deployment in support — including, crucially, support for the self-hosted and private deployments themselves — so that an organization running Mistral on its own infrastructure is not alone with it. This is a large part of what an enterprise is actually buying: not just the right to run the model, but a partner committed to keeping it running.

Data handling and training guarantees. One more piece belongs in this list because it is the question regulated buyers ask earliest: what happens to the data an organization's employees put into the tool? Enterprise deployment is where the answers a serious buyer needs become contractual rather than hopeful — that business data is not used to train models, that retention is bounded and specified, that the handling matches the residency commitments of Chapter 16. For an individual on a free plan these terms are something to read; for an enterprise they are something to negotiate and verify, and the enterprise offering exists partly so that there is a counterparty able to make those commitments in a form a legal team can accept. (Confirm the current specifics against Mistral's enterprise terms — exact retention and training language is the kind of thing that evolves.)

Taken together, these are the things that turn a capable model into deployable enterprise software. The capability was never the hard part; *this* is.

To see why, it helps to follow a rollout that does *not* have these pieces and watch where it breaks. A team enthusiastic about AI starts by having people sign up for individual accounts and expense them. For a while it looks like progress — usage climbs, people are getting value. Then the questions arrive, in roughly this order. Security asks who has access and cannot get a straight answer, because access is scattered across personal accounts nobody is tracking. Legal asks where the data is going and discovers that different employees, on different plans, have agreed to different terms. An employee leaves and their account, with whatever they put into it, simply continues to exist. Finance notices it is paying for dozens of overlapping subscriptions with no central view. None of these problems is about whether the model is good; every one of them is about the absence of the administrative, identity, security, and contractual layer that the enterprise offering provides. The grassroots rollout does not fail because the AI was weak. It fails because "a lot of individuals using a capable tool" is not the same thing as "an organization running a system," and the gap between those two is precisely what "enterprise" names.

How an Organization Chooses

Faced with a spectrum rather than a single option, the natural question is how to pick. The honest method is to start not with the technology but with the constraints, and let them narrow the field.

Begin with the residency and data requirements, because they are usually the hardest line and they eliminate options fastest. If your most sensitive data legally cannot leave your premises, that single fact points you toward on-premises or self-hosted for that data, and the rest of the decision follows from there. If your requirement is EU residency under EU law but third-party processing is acceptable, the hosted EU or private-cloud options are open to you. Let the strictest constraint set the floor.

Then weigh your operational capacity honestly. Running models on your own infrastructure demands hardware, expertise, and the willingness to own uptime and security — the burden of Chapter 15, now at organizational scale. If you have a capable platform team, on-premises is viable; if you do not, private cloud or hosted gives you most of the control with far less of the weight, and pretending otherwise leads to a deployment nobody can keep running. Match the deployment to the people and infrastructure you actually have, not the ones you wish you had.

Finally, recognize that the answer can be plural. A large organization does not have to make one choice for everything. The realistic pattern is to place the most sensitive workloads at the high-control end and the routine ones where they are cheapest and easiest to run, using Mistral's spectrum to put each workload at the point that fits it. The flexibility to do exactly that — to not be forced into one shape for all of your data — is, for many enterprises, the deciding advantage over a provider that offers only its own cloud.

The Sovereignty Proposition, Made Real

Step back and Vibe Enterprise is the whole argument of this book, finally assembled into something an organization can sign for. Chapter 14 established that the weights are genuinely open. Chapter 15 showed that you can run them yourself. Chapter 16 explained why, for a great many institutions, running them under your own control is not a preference but a requirement. This chapter is where those threads converge into a product: a frontier assistant, deployable from shared cloud all the way to your own data center, wrapped in the administration, identity integration, security, and support that an institution needs before it can put a tool in front of its whole workforce.

That is the sovereignty proposition stated in full. The American flagships offer capability on their terms. Mistral, at the enterprise level, offers comparable capability on *your* terms — where the data lives, who can reach it, which laws govern it, and who answers the phone when something breaks, all chosen by you rather than dictated to you. For the manufacturer that lost a quarter to the unglamorous questions, that is not a marginal advantage. It is the difference between an AI rollout that clears legal, satisfies security, integrates with identity, and actually ships — and one that stalls forever on the stack of questions no demo ever answers.

With that, Part V closes. We have moved from the principle of open weights to the practice of running them, from the *why* of sovereignty to the *how* of deploying it across an organization. Part VI turns to the developer's side of Mistral — La Plateforme, the API, model choice, fine-tuning, and the economics of building on these models — where the same openness that made self-hosting possible becomes the foundation for building software of your own.

Part VI: Building on Mistral

Audience: developers and builders. Readable to the technically curious; navigate around the deep parts if code isn't your work.

"The spectrum is the product: hosted API, private cloud, or fully self-hosted. You pick the point where capability meets the control you actually need."

— A working principle

Chapter 18: La Plateforme: The API

Every chat-only workflow eventually hits the same wall. Picture a developer who built a clever little tool inside the Vibe chat box — a prompt that took a customer's support email, classified its urgency, drafted a reply, and flagged anything that needed a human. It worked beautifully, and then the wall arrived: he had to paste each email in by hand and copy each answer out, one at a time, like a very smart intern who could only be handed work on paper. What he actually needed was for his own software to do the handing — to take the thousand emails already sitting in his support inbox and run every one through Mistral without a person in the loop. The capability was right; the doorway was wrong. He needed the API. This chapter is about that doorway.

Everything in this Part is written for people who build — developers wiring Mistral's models into their own software. If you only ever use Mistral through Vibe, you can skip Part VI without losing the thread; the rest of the book is yours. But it is worth knowing this surface exists, because it is where Mistral stops being an assistant you open and becomes a capability you build with. And one caution belongs at the very top, the same one Chapter 4 raised about model names: API details — endpoints, parameter names, exact rates, limits — are the most perishable material in the book. I will describe the durable shape and flag every specific. **The authority is always Mistral's current API documentation, not this page.**

What La Plateforme Actually Is

La Plateforme is Mistral's API and developer platform — the programmable way to reach the same models that power Vibe. Instead of a person typing into the Vibe interface (Chapter 6) and reading the reply, your program sends Mistral a request over the network — a prompt, some context, some settings — and receives the model's response back as structured data your code can use. The intelligence is the same family of models. What changes is who does the asking. In Vibe, a human is in the loop. Through La Plateforme, your software is.

That distinction is the whole point. It means Mistral's capabilities — the reasoning, the writing, the analysis, the code generation — can live *inside* applications you write: a support system that drafts replies, a pipeline that summarizes documents as they arrive, an internal tool that classifies and routes on demand. Vibe is Mistral for people. La Plateforme is Mistral for programs. And because Mistral's distinctive trait is openness (Part V), La Plateforme is only one of several ways to run these models — the convenient, hosted, pay-as-you-go end of a spectrum that runs all the way out to weights you host yourself. Chapter 21 maps that full spectrum; this chapter is about the hosted-API end of it.

The shift from "person in the loop" to "program in the loop" is larger than it first sounds, and it is worth dwelling on because almost everything else in this Part follows from it. When a person uses Vibe, they are a built-in safety system: they read each answer before acting on it, they notice when the model has misunderstood, they catch the obvious mistake, and they simply stop when something looks wrong. The API removes that person. A program does not read the answer with judgment; it does whatever it was told to do with the output, at whatever scale and speed you set it running. That is the source of the API's power — a thousand emails handled while you sleep — and the source of its hazard in equal measure, because the same loop that processes a thousand good results will process a thousand bad ones just as fast if something upstream goes wrong. The developer's hand-pasting workflow was slow and tedious, but the tedium was also a brake: he saw every input and every output. Automating it removes the brake along with the tedium, and the responsibility for replacing that brake with something deliberate — validation, limits, monitoring — moves to you. Hold that thought; it is the throughline of the whole chapter.

Who It's For

Three rough groups reach for La Plateforme, in my experience.

Builders embedding Mistral in a product — a team or a company putting AI into an application, where a Mistral model is the engine behind a feature their users touch. This is the largest group and the one this Part is mainly written for.

Automators and integrators — people wiring Mistral into pipelines, batch jobs, and back-office tooling, where there is no user-facing chat at all, just a model doing a step inside some larger automated process: tagging records, extracting fields, summarizing on a schedule.

Evaluators and prototypers — developers comparing Mistral against other providers, or testing whether an open-weight, European-hosted model changes what they are allowed to build. For organizations with data-residency constraints (Chapter 16, Chapter 23), the API hosted on European infrastructure is often the first serious foot in the door, before any conversation about self-hosting.

If none of those is you, Part VII is a clean place to rejoin the book. If one of them is, read on.

Getting Oriented: Keys, Auth, and Endpoints

The mechanics of using La Plateforme follow a shape you may already recognize from other AI providers, and I will keep this conceptual on purpose, because the precise strings are exactly what drifts.

You authenticate with an API key. Programmatic access uses a key — a secret credential you generate from Mistral's developer console, attached to each request, rather than the interactive sign-in a human uses for Vibe. The key identifies your account, meters your usage, and is the thing you must guard. Treat it like any other secret: store it in environment variables or a secrets manager, never paste it into client-side code or commit it to a repository, and rotate it if it leaks. A leaked key is someone else spending your money against your account. **Generate and manage keys from Mistral's current developer console, and confirm the exact setup steps there.**

You send requests to an endpoint. Your code makes network calls to Mistral's API — a base URL with paths for the operations it supports, the central one being a chat-style request where you send a list of messages and get the model's reply back. The request carries your key, the model you want, your messages, and settings like a cap on the reply length. **The exact base URL, paths, and parameter names live in Mistral's docs and are check-the-docs details — do not hardcode them from memory or from this book.**

The shape will feel familiar. La Plateforme's API is broadly industry-standard in design — a chat-completions style of interaction that developers coming from other AI providers tend to find recognizable, sometimes to the point where existing client libraries or patterns carry over with little change. I want to be careful here, because "compatible" is a word that gets oversold: the degree of drop-in compatibility, which libraries work, and which features map cleanly are precisely the things that shift release to release. So hold the durable claim — *the API follows familiar, conventional patterns, and Mistral provides its own official client libraries* — and verify the specific compatibility you are counting on against the current documentation rather than trusting a blanket promise.

You handle failure as a normal case, not an exception. This is the part newcomers most often skip and most often regret. A network call to a model can fail in ordinary ways that have nothing to do with your code being wrong: the request can time out, the service can be briefly unavailable, you can hit a rate limit because you are sending requests faster than your account is allowed, or a particular input can be rejected. In Vibe, a person simply tries again. In your software, *you* are the one who has to decide what "try again" means — whether to retry after a short, increasing wait, how many times before giving up, what to show the user or write to a log when a call ultimately fails, and how to make sure a burst of retries does not itself become the thing that trips the rate limit. None of this is exotic; it is the same defensive plumbing any networked service needs. But because the happy path is so easy to get working in a demo, the failure handling is what separates a prototype from something you can leave running unattended, and it is worth building in from the first real version rather than bolting on after the first outage.

Rate limits and quotas shape how fast you can go. Related to failure handling but worth naming on its own: your account has limits on how many requests, or how much total work, it can send in a given window, and those limits exist to protect the shared service as much as your bill. A batch job that fires a thousand requests as fast as it can will likely be throttled; the well-behaved version paces itself, runs requests in controlled concurrency rather than all at once, and treats a rate-limit response as a signal to slow down rather than an error to hammer through. The exact limits, and how to request higher ones, are account-specific and live in the console — but the discipline of designing your workload to live within a rate limit, rather than discovering it in production, is universal.

A Minimal Illustrative Request

To make this concrete, here is the *shape* of a basic request in pseudocode. I am writing it loosely on purpose — this is meant to show you the moving parts, not to copy-paste. The real parameter names, the real endpoint, and the real client library belong to Mistral's current docs, and Appendix B walks through a working quickstart you can actually run.

```
# ILLUSTRATIVE ONLY – verify every name against current Mistral API docs
POST  <Mistral chat endpoint>
Header:  Authorization: Bearer <YOUR_API_KEY>
Body:
  model:     "<current model name>"   # e.g. a Mistral Large or Medium model
  messages:  [
    { role: "system", content: "You are a helpful assistant." },
    { role: "user",   content: "Summarize this support email: ..." }
  ]
  max_tokens: <a cap on the reply length>

# Response comes back as structured data containing the model's reply.
```

Four things to notice, because they are durable even as the names drift. You pick a **model** — which member of the family answers, the decision Chapter 19 is entirely about. You send a list of **messages** with roles, typically a system message that sets behavior and a user message that asks. You set **limits** like a maximum reply length. And you get back **structured data**, not loose text, so your program can act on it. Every modern AI API is some version of this, which is the good news: if you have used one, La Plateforme will not feel foreign.

It is worth tracing that developer's support-email tool through this shape end to end, because it shows how a chat-box experiment becomes a real piece of software. In Vibe, his whole tool was a single prompt he pasted an email into. Through the API, that same logic becomes a small program with a few deliberate parts. First, the program *reads* the emails from wherever they live — the inbox, a database, a queue — one at a time or in controlled batches. For each one it *builds a request*: a system message that fixes the model's job ("You classify support emails by urgency and draft a first-pass reply; respond in this exact structure"), and a user message carrying the email text. It *sends* that request, *handles the failures* discussed above — a timeout here, a rate limit there — and then *parses* the structured reply, because the model's output is data the program now has to act on rather than text a human will read. Crucially, it does not blindly trust that output: it *validates* it (is the urgency one of the allowed values? did a draft actually come back?) before doing anything consequential with it. And because some emails will be genuinely ambiguous or sensitive, it *routes the uncertain ones to a human* rather than sending an automated reply into a situation that needs judgment. The capability at the center is identical to the chat-box version;

everything wrapped around it — reading, building, sending, retrying, parsing, validating, escalating — is the work of turning a clever prompt into software you can leave running. That wrapper, not the prompt, is where the real engineering lives, and it is the same wrapper whether the model behind it is Mistral's or anyone else's.

Two Things That Genuinely Differ from Vibe

It helps to be clear about what the API is and is not relative to the Vibe assistant (Chapter 6). Same model family, different doorway — but two things genuinely differ, and both matter.

The API is billed separately from your Vibe subscription, and there are no shared credits. This is the single most important practical fact in the chapter, so I will say it plainly: paying for Vibe as a person — even an enterprise Vibe plan (Chapter 17) — does not cover programmatic API usage, and a Vibe subscription does not top up an API balance or the other way around. La Plateforme bills on its own pay-per-use meter, by the token, against its own account and its own payment method. They are two separate commercial relationships that happen to share a model family. Know which meter you are on before you build something that runs often; Chapter 21 is the honest accounting of what it costs.

The API hands you fewer guardrails and more responsibility. Vibe wraps the model in an interface, a history, and a set of safety and rate behaviors built for humans. The API hands you the raw capability and expects you to build the surrounding judgment — the input validation, the output checking, the cost controls, the guardrails your own users will need. That is power and obligation in the same package, and Part VII's responsible-use material applies with extra force the moment a model is acting inside software rather than in front of a person reading every reply.

What the API does *not* change is the model's character: its strengths, its failure modes, its tone are the same through La Plateforme as in the Vibe box. What it adds is reach — your software, not just your fingers, can now put the model to work.

So that is the doorway. The next decision, and the one that most shapes both your bill and your results, is which model walks through it — because the API exposes the whole family, and they are not interchangeable. And for the hands-on, run-it-yourself version of everything sketched here — a key, a real request, a first response — Appendix B is the quickstart to keep open beside the docs.

Chapter 19: Choosing a Model for the Job

Choosing a model is not picking the smartest one. It is matching the model to the task — and the most expensive mistake in building on an API is to skip that match and wire the most capable model into every call. Consider the team whose first real bill arrived three times higher than the estimate, for a reason almost embarrassing in its simplicity: they had pointed the flagship at every request, including the ones that did nothing harder than decide whether an incoming message was a question or a complaint. They were paying flagship rates for work a far smaller model could have done in a fraction of the time and cost. Nobody had made a *bad* decision. They had made one decision — "use the best model" — and applied it everywhere, when the job in front of them was a dozen different jobs of wildly different difficulty. This chapter is about making that match deliberately.

The good news with Mistral is that there is a genuine family to choose from, and — because of the open-weight story — an unusually wide one, from large flagship models down to small models you could run on a laptop (Chapter 4, Appendix A). The bad news is that the specific roster moves fast, so I will name the current members as reference points and ask you to confirm them. **The exact model names, their capabilities, and their rates are check-the-docs details; the authority is Mistral's current model documentation, not this page.**

The Three Tiers, Conceptually

Set the exact names aside for a moment and think in tiers, because the tiers are durable even as the members shuffle.

The flagship tier — the most capable models, for the hardest work. As of this writing this is **Mistral Large 3** (released under the permissive Apache 2.0 license, the most capable in the lineup — **verify**). You reach for the top tier when the task needs the strongest reasoning, the most reliable structured output, the subtle judgment, the long and tangled context. It costs the most per token and is, for the same input, the slowest to respond. You pay that premium where the difficulty actually lives.

The efficient mid tier — strong capability at much lower cost and latency. As of this writing this is **Mistral Medium 3.5** (under a modified-MIT license, and the engine behind much of Vibe — **verify**). This tier is the workhorse: capable enough for the large majority of real tasks, meaningfully cheaper and faster than the flagship, and very often the right *default* rather than a compromise. A surprising number of teams who think they need the flagship are actually well served here.

The small-and-fast tier — for volume, simplicity, and tight latency. Smaller models, including ones light enough to run on modest hardware yourself, handle the high-volume, low-difficulty work — classification, routing, extraction, simple drafting — at the lowest cost and the quickest response. The task that decided "question or complaint?" in the opening story belongs here, not at the top.

Hold those three tiers as the stable structure. The specific model that occupies each tier, and the rates attached, are the perishable part — look them up.

The cleanest way to build intuition for the tiers is to sort a handful of real tasks into them and notice the pattern. *Decide whether an incoming message is a question, a complaint, or spam* — small tier; it is a narrow judgment on short text, run constantly, where speed and cost matter far more than brilliance. *Pull the invoice number, date, and total out of a scanned document* — small tier again; extraction against a known structure is exactly what the cheapest models do well. *Draft a friendly first-pass reply to a routine customer email* — mid tier; it needs real fluency and a sense of tone, but it is not hard reasoning, and the mid tier produces drafts that need only light editing. *Read a forty-page contract and explain which three clauses expose us to the most risk and why* — flagship; this is subtle, multi-step reasoning over long, tangled context, the kind of work where the top tier's extra headroom actually earns its premium. Lay those four side by side and the lesson is hard to miss: they are not one job at four difficulties, they are four genuinely different jobs, and pricing them all at flagship rates because they all happen to involve "AI" is the precise error the opening team made.

The Four Dimensions You Are Trading Off

Picking a tier is really a trade-off across four things, and naming them keeps the decision honest.

Capability. Can the model actually do the task well and reliably? This is the dimension everyone starts with and the one that, past a certain point, stops mattering — because for a great many tasks the mid tier is already more than capable enough, and the flagship's extra headroom is paying for ability you are not using.

Cost. What does each call cost, multiplied by how often your software makes it? At human scale, in the Vibe box, model cost is mostly invisible. At software scale, where your code makes the calls, it is the dominant line item, and it scales directly with how often you run and which tier you run. Chapter 21 is the full accounting; here it is enough to say that the gap between tiers, multiplied by volume, is usually the largest number in the project.

Latency. How fast does the answer come back? A bigger model generally takes longer to produce the same reply. For a batch job that runs overnight, nobody cares. For a feature where a user is waiting on the result, a slower model is a worse product even if its answers are marginally better. Match the tier to how much the wait matters.

Openness. This one is Mistral's distinctive dimension, and it does not exist on a closed provider's menu at all. Because some models are released as open weights under permissive licenses, your choice of model also chooses your *future options*: pick an open-weight model and you keep the door open to fine-tuning it yourself (Chapter 20) or moving it off the hosted API onto your own infrastructure (Chapter 15, Chapter 21) without re-platforming. Two models that look equally capable today are not equivalent if one locks you into the hosted API and the other can follow you wherever you need to run it. For sovereignty-conscious work, openness can outweigh a small capability gap entirely — which is the whole argument of this book, now showing up as a line in a config file.

The Heuristic: Start Cheaper, Escalate If Needed

Here is the working discipline, and it inverts the instinct that cost the opening team their tripled bill. **Do not default to the most capable model. Default to the cheapest tier that plausibly does the job, test it honestly, and escalate only where the results actually fall short.**

In practice that means starting a new task on the small or mid tier, not the flagship, and looking at the real output on real inputs. If a small model classifies your support emails correctly, you are done — escalating to the flagship buys you nothing but a larger bill. If the mid tier drafts replies that need only light editing, you are done there too. You move up a tier only when you can point to a specific way the cheaper model is failing the task: missing nuance, getting structure wrong, breaking on the hard cases. Capability you cannot demonstrate a need for is capability you are paying for and not using.

This escalate-only-on-evidence approach has a second payoff beyond cost. It forces you to actually look at output quality rather than assuming the biggest model launders away every problem — and it surfaces the cases where the issue is your prompt or your data, not the model, which no amount of escalation would have fixed.

To make the discipline operational rather than aspirational, it helps to do the testing properly, which mostly means resisting the urge to judge a model from one or two impressive examples. Assemble a small set of real inputs — twenty or thirty is often plenty to start — and deliberately include the awkward ones: the ambiguous email, the malformed document, the edge case that is rare but expensive when it goes wrong. Run that set through the cheaper tier and *read the outputs yourself*, ideally against what a correct answer would have been. A model that handles your twenty-five easy cases and fails the five hard ones has told you something precise: not "this model is bad" but "this model needs a fallback for this specific kind of input," which often means routing just those cases up a tier rather than moving everything up. The teams that get the most out of a cheap tier are the ones who looked closely enough to know exactly where it breaks, because that knowledge is what lets them escalate surgically instead of wholesale.

A related trap is worth flagging because it quietly defeats the whole exercise: when a cheaper model disappoints, the issue is frequently not the model's capability but the instructions it was given. A classification that comes back inconsistent often improves dramatically when the prompt spells out the exact categories and what each one means; a draft that misses the mark often just needed an example of the desired tone. Escalating to a more expensive model can paper over a vague prompt — the bigger model guesses your intent better — but you will be paying flagship rates to compensate for instructions a small model could have followed if they had been clear. Before you conclude the cheap tier cannot do the job, make sure you have actually told it what the job is. The cheapest capability upgrade in any build is usually a better prompt, not a bigger model.

Routing: Different Models for Different Calls

The most cost-effective applications rarely use one model for everything. They **route** — sending each call to the tier that call deserves. The cheap classifier decides what kind of request came in; the mid tier handles the bulk of the real work; the flagship is reserved for the genuinely hard cases the cheaper models flagged or fumbled. One application, three tiers, each doing the work it is suited to.

The support-email pipeline from the previous chapter is the canonical example of routing done well. Every incoming email first hits a small, cheap model whose only job is to sort it: routine question, angry complaint, billing dispute, spam. That classification — fast and nearly free — then decides where the email goes next. The routine questions, the bulk of the volume, go to the mid tier for a drafted reply. The genuinely fraught ones — the legal threat, the safety issue, the case the classifier itself flagged as uncertain — are escalated to the flagship or handed to a human. The result is that the expensive capability is spent only on the small slice of traffic that actually needs it, while the cheap tier absorbs the volume. The same total workload that would have cost flagship rates on every call now costs flagship rates on a fraction of them, with no loss in quality where quality matters, because the hard cases still reach the strong model.

There is a subtler design point hiding in that example. Routing is not only a cost lever; it is also a *reliability* lever, because the cheap classifier doubles as a triage step that decides what deserves human attention. A well-designed pipeline does not just route by predicted difficulty — it routes by predicted *risk*, sending anything sensitive or ambiguous to a person regardless of how confident the model is. That instinct, to let the cheap model decide what the system should not handle automatically at all, is often more valuable than the cost savings, because it is the mechanism that keeps an automated pipeline from confidently mishandling the one case that should never have been automated.

You do not have to build elaborate routing on day one — and for a simple tool, a single well-chosen mid-tier model is a perfectly good answer. But knowing that routing is available keeps you from the trap the opening team fell into, where one global "use the best" decision quietly multiplies the bill across every call that never needed it. The largest cost lever in any Mistral build is which model answers which request, and routing is how you pull that lever at the level of the individual call rather than the whole application.

When the Decision Is Already Made for You

Two cases short-circuit all of this, and they are worth naming so you do not overthink them.

If your constraint is **data residency or self-hosting** (Chapter 16, Chapter 23), your choice narrows to the open-weight models you are permitted to run where you need to run them, and openness moves from a tie-breaker to the first filter. The capability comparison happens *within* that permitted set, not across the whole catalog.

And if you are simply **prototyping**, do not agonize. Reach for a capable mid-tier model, get the thing working end to end, and tune the model choice once you can see real usage and real cost. Optimizing model selection before you know your actual traffic is optimizing a number you have not measured.

With a model chosen and a job in hand, the next question is whether the stock model is enough — or whether teaching it your specific domain, your data, your house style would earn its keep. That is the question fine-tuning answers.

Chapter 20: Fine-Tuning and Customization

Fine-tuning is a genuinely powerful tool, and it is also the tool people reach for first when they should reach for it last. Consider the legal-tech team that decided, with some pride, to fine-tune a model — and when you asked what problem they were solving, the answer was that the model did not reliably format citations the way their firm liked. That is a real annoyance, but it is not a fine-tuning problem; it is a one-paragraph instruction in the prompt. They were about to spend weeks and a meaningful budget assembling a training dataset to fix something a well-written example in the system message solved in an afternoon. This chapter is about what fine-tuning actually is, when it earns its considerable cost, and when something cheaper does the job better.

As always with this Part, the durable ideas are stable and the specifics are not. **How fine-tuning is offered on La Plateforme, what it costs, which models support it, and the exact mechanics are all check-the-docs details; the authority is Mistral's current documentation, not this page.**

What Fine-Tuning Actually Is

Fine-tuning takes an existing trained model and adjusts its weights further on examples you provide, so the model internalizes a pattern, a style, a domain, or a behavior that you want it to produce reliably without being told every time. The stock model already knows language and reasoning broadly; fine-tuning bends that general competence toward your particular world. You feed it many examples of the input-and-ideal-output you care about, and it comes out the other side biased toward producing that kind of output by default.

The result is a model that *behaves* like your examples without needing the instructions and demonstrations spelled out in every prompt. A model fine-tuned on your support team's best replies tends to write in that voice on its own. A model fine-tuned on your domain's terminology stops fumbling the vocabulary. The knowledge or the style moves from something you supply at request time into something baked into the weights.

That is also the clearest way to understand its limits. Fine-tuning is good at teaching a model *how to behave* — tone, format, style, the shape of a good answer in your domain. It is far weaker, and often the wrong tool, for teaching a model *new facts* it needs to look up, especially facts that change. For that, you usually want a different technique entirely.

It helps to be concrete about the mechanics, because the word "training" makes people imagine something grander than what actually happens. A fine-tuning example is, in the common case, a short conversation: a user turn, and the assistant turn you wish the model had produced. You assemble a few hundred or a few thousand of these into a file — typically one example per line in a structured format — and the training process walks through them repeatedly, nudging the weights a little at a time so that, on inputs like the ones in your examples, the model's natural output drifts toward the outputs you supplied. Nothing about the model's general intelligence is replaced; it is the *defaults* that shift. Ask a fine-tuned support model an unrelated trivia question and it still answers like the base model, because you never gave it examples that pulled trivia in a new direction. This is why fine-tuning is so good at consistency and so poor at fact injection: it changes habits, not knowledge.

The Three Tools, and Why Order Matters

Before fine-tuning anything, it is worth knowing the cheaper tools it competes with, because reaching for fine-tuning first is the most common and most expensive mistake.

Prompting comes first. A great many problems that look like they need a fine-tuned model are really just under-specified prompts. Clear instructions, a few well-chosen examples shown right in the prompt, a system message that sets the role and the rules — these solve more than people expect, including my legal team's citation formatting. Prompting costs nothing but thought, changes instantly when you change your mind, and should always be the first thing you exhaust.

Retrieval comes second. When the real problem is that the model needs to *know things* — your documents, your product details, your current policies — the right tool is usually retrieval, often called RAG: you fetch the relevant information at request time and hand it to the model alongside the question, so it answers from your actual content rather than from training. Retrieval is how you give a model facts, especially facts that change, without retraining anything. If your problem is "the model does not know about our internal handbook," that is a retrieval problem, not a fine-tuning one, and trying to fine-tune the handbook into the weights is both harder and worse — it goes stale the moment the handbook updates.

Fine-tuning comes third — for what the first two cannot reach. You earn your way to it after prompting and retrieval have been tried and found wanting, not before.

The order is not arbitrary; it tracks cost, reversibility, and how fast you can iterate. Prompting is free to change and you see the effect in seconds, so you can try ten variations over a coffee. Retrieval costs some engineering to set up but, once built, updates the moment you add a document, with no retraining and nothing baked in to go stale. Fine-tuning sits at the far end: it costs money and time to produce, it freezes a behavior into weights that only another training run can change, and you cannot tell whether it worked until you have evaluated it. Working up the ladder in that order means you never pay the expensive, slow, hard-to-reverse cost until you have proven the cheap, fast, reversible options genuinely cannot do the job. The teams that skip to step three first almost always discover, months later, that step one would have sufficed.

A worked example of the order doing its job: a fintech company wanted an assistant that answered customer questions about their loan products in the company's careful, hedged, legally reviewed voice, using the current rates and terms. They nearly fine-tuned the whole thing. Walked up the ladder instead, they found the voice was a *prompting* problem — three good example exchanges in the system message and a paragraph of tone rules got them most of the way. The current rates and terms were a *retrieval* problem — those change weekly and must never be wrong, so they are fetched live from the product database at request time, never trained in. What was left for fine-tuning, if anything, was a subtle and consistent stylistic discipline that the prompt examples almost-but-not-quite held across thousands of calls. They ended up fine-tuning a thin sliver of the problem, having solved the bulk of it for almost nothing.

When Fine-Tuning Actually Wins

So when is it the right tool? A handful of situations where it genuinely beats the cheaper options.

When you need a consistent style or format that examples-in-the-prompt cannot reliably hold. If you need the model to produce output in a very specific voice, structure, or convention every single time, across thousands of calls, and stuffing examples into every prompt is either unreliable or wasteful, fine-tuning bakes that consistency into the model itself.

When prompt overhead is itself a cost. Long prompts full of instructions and examples cost tokens on every call (Chapter 21). If the same heavy instructions ride along on millions of requests, fine-tuning that behavior into the model and shrinking the prompt can pay for itself in token savings at scale — a smaller, cheaper model fine-tuned for your task can sometimes outperform a larger stock model on that task, which is a real cost win on both axes.

When you have a genuinely specialized domain the stock model handles awkwardly. Niche terminology, an unusual task structure, a domain where the general model is competent but not fluent — these are where fine-tuning earns its keep, teaching the model the patterns of your world more deeply than any prompt can.

When you have the data, and it is good. Fine-tuning lives or dies on the quality of your examples. A few hundred to a few thousand clean, representative, correct examples is the kind of thing that works; a pile of noisy, inconsistent, or wrong examples teaches the model to be noisy, inconsistent, and wrong. If you do not have good labeled examples and cannot reasonably get them, you do not yet have a fine-tuning project — you have a data project that comes first.

It is worth making these abstract wins concrete with cases where the math actually favored fine-tuning. A medical-coding company needed a model that read clinical notes and emitted billing codes in a rigid, structured format with extremely specific conventions — the kind of task where a small deviation breaks a downstream system. Prompt examples got them to roughly the right shape but not to the reliability they needed across the volume they ran, and the instruction block required to push reliability higher was so long it dominated the cost of every call. Fine-tuning on a few thousand verified note-to-code pairs solved both at once: the structured output became near-perfectly consistent, and because the conventions were now in the weights, they could strip the giant instruction block out of the prompt and shrink their per-call cost substantially. That is the textbook win — consistency at scale *and* a token saving, on a specialized domain, with good data they already had from years of human coders.

A second case shows the same engine from a different angle. A localization team translated marketing copy into a house dialect of several languages — not just "French," but their French, with brand-specific terms left untranslated, a particular register, and conventions a generic translator kept flattening. No amount of prompting reliably held the house style across the breadth of content they pushed through. A model fine-tuned on their translation memory — years of human-approved translations in exactly the voice they wanted — learned the house dialect as a default, and the editors' correction load dropped sharply. The data existed because the work had been done by humans for years; the fine-tune simply distilled that accumulated judgment into the model.

Two Roads: La Plateforme, or Do It Yourself

Here is where Mistral's openness opens a door closed providers leave shut. There are two genuinely different ways to fine-tune, and which one fits depends on the same control-versus-convenience axis that runs through this whole book.

Fine-tune through La Plateforme. Mistral offers managed fine-tuning on its platform: you supply your examples, Mistral runs the training, and you get a customized model you call through the same API as any other (Chapter 18). This is the convenient road — no infrastructure, no machine-learning operations of your own, and the result is a hosted model you query like the stock ones. The trade-offs are the usual hosted ones: you are on the provider's platform, your training data goes to Mistral to do the training, and you pay Mistral's pricing for the work. For most teams who want a tuned model without building a training pipeline, this is the right road. **Confirm current availability, supported models, and pricing in Mistral's documentation.**

Fine-tune the open weights yourself. Because Mistral publishes the weights of its open models (Chapter 14), you can also do the fine-tuning entirely on your own infrastructure, with your own tooling, and never send your training data anywhere. This is the road that exists *only* because of the open-weight model, and it is the one that matters for sovereignty-bound work (Chapter 16, Chapter 23): the sensitive data you would train on — exactly the data you are not allowed to export — never leaves your control, and the resulting tuned model is yours to run where you run everything else (Chapter 15). The cost is that you take on the whole burden: the hardware, the machine-learning expertise, the operational care of doing it well. It is more capable and more demanding in equal measure.

The do-it-yourself road has a further subtlety worth naming, because it changes the cost picture more than people expect. Modern self-hosted fine-tuning rarely means retraining the whole model from its billions of parameters; the common practice is a lightweight technique that trains a small set of additional parameters layered onto the frozen base model. The practical upshot is that fine-tuning the open weights on your own hardware is far more attainable than "retrain a frontier model" sounds — it can fit on hardware a serious team already owns, the training runs in hours rather than weeks, and the small trained layer is a compact artifact you can version, swap, and keep alongside the base weights. This is the quiet reason self-hosted fine-tuning is a real option for ordinary organizations and not just for labs: the open weights plus a lightweight method bring the sovereign road within reach of teams who could never afford to train from scratch. The expertise barrier is real and the operational care is real, but the compute barrier is lower than the framing suggests.

The choice between the two roads is not really about fine-tuning at all. It is the deployment-spectrum choice from the rest of this Part, seen from the training side: convenience and a hosted result on one end, total control and total responsibility on the other, with the open weights being what makes the second end possible.

A Realistic View of Effort and Payoff

I will close with the honest accounting, because fine-tuning is the customization technique most prone to being romanticized. It is real work. There is data to assemble and clean, training runs to manage and pay for, evaluation to do so you actually know the tuned model is better and not just different, and a tuned model to maintain as your needs drift. None of that is free, and a fine-tuned model you do not evaluate is a liability dressed as an upgrade — you have changed the model's behavior without confirming you changed it for the better.

So the durable verdict is the cautious one. Reach for fine-tuning when prompting and retrieval have genuinely come up short, when you have good data, and when the payoff — consistency at scale, lower per-call cost, real domain fluency — is large enough to clear the cost of getting and keeping it. When those conditions hold, fine-tuning is a strong tool and, on open weights, a sovereign one. When they do not, the cheaper tools you skipped past were the right answer all along, and the citation-formatting team is your cautionary tale.

Whichever road you take, the bill eventually comes due — and the shape of that bill depends enormously on where and how you run the model, across the full deployment spectrum and what each point on it costs.

Chapter 21: Cost and Deployment Options

Two companies I advised in the same quarter ran almost identical workloads on Mistral models and arrived at almost opposite right answers about how to pay for it. One was a small team with spiky, unpredictable traffic — busy some weeks, quiet others — and for them the hosted API was obviously correct: pay only for what you use, nothing to run when nobody is asking. The other processed an enormous, steady river of documents around the clock, the same heavy load every hour of every day, and for them the per-token meter would have been a slow financial bleed; self-hosting the open weights on their own hardware cost a large fixed sum and then ran nearly free per request, which at their volume won by a wide margin. Same models, same kind of work, two correct and contradictory decisions — because deployment is not one question with one answer. It is a spectrum, and where you belong on it depends on your volume, your data rules, and what your team can operate. This chapter is the map.

The discipline here is the one this whole Part has insisted on, and it matters most where money is involved. **The structure of the deployment spectrum and the cost models is durable. Every specific rate is not, and the only authority on today's figures is Mistral's current pricing — check it before you budget anything against it.** I will give you reference points of the right order of magnitude and flag each one. Treat them as the shape of the thing, never as a quote.

The Deployment Spectrum

Mistral's distinctive offering is not a single product but a *range* of ways to run the same models, and the range itself is the point — most providers give you one option, their cloud on their terms (Chapter 1). Picture three stations along a line from most convenient to most controlled.

The hosted API — La Plateforme. Mistral runs the models on its own European infrastructure; you reach them over the network and pay per token for what you use (Chapter 18). Nothing to provision, nothing to operate, instant to start, and it scales itself. Your data passes through Mistral's systems to be processed, and you pay a per-use meter. This is the convenient end, and for most teams most of the time it is the right starting point.

Private cloud and dedicated deployment. A middle station where Mistral's models run on infrastructure dedicated to you — in a private cloud, or a managed deployment that keeps your workload isolated rather than sharing the public API (Chapter 17). You gain stronger data-handling and residency guarantees and more predictable capacity; you take on more setup and usually a different, more committed cost arrangement. This is where many regulated organizations land: more control than the public API, less operational burden than running everything yourself.

Fully self-hosted. The far end, and the one only Mistral's openness makes possible (Chapter 14, Chapter 15). You download the open weights and run the model on infrastructure you control — your servers, your private cloud, your own hardware — with nothing transmitted to anyone. You pay for compute and operate the whole thing yourself. This is maximum control and maximum responsibility, and for sovereignty-bound work where the data simply cannot leave (Chapter 16, Chapter 23), it is often the only station that clears the bar at all.

The crucial thing is that these are points on a continuum you can move along, not locked-in products. A common path is to prototype on the hosted API, prove the workload, and migrate toward private or self-hosted deployment as volume and data requirements grow — and because the open-weight models are the same models, that migration does not mean re-platforming onto a different vendor.

It is worth dwelling on why that last point matters more than it first appears. When you prototype on a *closed* provider's API and later decide you need to bring the workload in-house, you cannot — there is no in-house version of a black box, so a move to self-hosting means abandoning the model entirely and re-validating everything against a different one. With Mistral, the model you tested on the hosted API and the model you can run on your own hardware are the same family of weights. You can validate cheaply and conveniently on the API, then, if your volume or your data rules push you toward self-hosting, carry the proven workload across without throwing away the validation work. The spectrum is not three separate products you choose between once; it is one capability you can relocate along the convenience-versus-control axis as your circumstances change.

Two Fundamentally Different Cost Models

The deepest thing to understand about that spectrum is that the two ends are not just cheaper and more expensive versions of the same bill. They are *different kinds of cost*, and conflating them is how budgets go wrong.

The hosted API is a per-token, pay-as-you-go cost. You pay for the input tokens you send and the output tokens the model generates, by the million (Chapter 19). As reference points for the current flagship — **and these are exactly the figures to verify** — Mistral Large 3 is in the neighborhood of **$0.50 per million input tokens and $1.50 per million output tokens (verify both)**, with output costing more than input, as is standard across the industry, and the more efficient Mistral Medium 3.5 priced lower still (**verify**). The defining feature of this model is that cost scales directly with usage: no traffic, no bill; ten times the traffic, ten times the cost. It is beautifully suited to spiky, uncertain, or growing workloads, because you never pay for capacity you are not using.

Self-hosting is a compute-and-operations cost, mostly fixed. When you run the open weights yourself, you are not paying per token at all — you are paying for the hardware (or the cloud compute you rent to host it), the people who keep it running, and the electricity and overhead underneath. That cost is largely fixed: it is roughly the same whether the model answers a thousand requests a day or a million, because you are paying to *have the capacity*, not to use it. The per-request cost can be very low once the fixed cost is paid, but the fixed cost is real and arrives whether or not you have the traffic to justify it.

That difference is the whole decision in miniature. Per-token billing punishes volume and rewards unpredictability; fixed compute punishes idleness and rewards steady scale. My two opening companies sat on opposite sides of exactly this line.

It pays to make the per-token meter concrete, because "by the million" hides how the bill actually accumulates. Tokens are roughly word-pieces; a useful rough rule, **worth confirming against current tokenizer behavior**, is that a token is around three-quarters of a word in English, so a thousand words is on the order of thirteen hundred tokens. Now picture a customer-support workflow where each request sends the model a system instruction, a few retrieved knowledge-base passages, and the customer's message — call it three thousand input tokens

— and the model writes a four-hundred-token reply. At the reference rates above, that single interaction costs a fraction of a cent. Trivial. But the lesson is in the multiplication, not the single call: ten thousand such interactions a day is a real daily figure, and a hundred thousand makes the input side — those retrieved passages riding along on every call — the dominant cost, not the model's answers. This is why people are surprised by their bills: the expensive part is usually not the output you can see but the context you stuffed in, invisibly, on every request. Trimming what you send is often the fastest saving available, which is why it returns as a discipline at the end of this chapter.

The fixed-cost side deserves the same concreteness. Self-hosting a capable open-weight model means standing up a server with enough accelerator memory to hold the model and serve requests, then keeping it healthy. Whether you buy that hardware outright or rent equivalent cloud compute by the hour, the meter runs the same whether the box is flat-out or idle at three in the morning. The arithmetic that decides everything is utilization: a self-hosted box you keep busy is cheap per request, and the same box answering a handful of requests a day is the most expensive way to run a model ever devised, because you are paying full freight for capacity you are barely touching. Self-hosting rewards a firehose and punishes a trickle, which is the precise inverse of the per-token meter.

How to Choose Your Station

Three questions settle it, and you should answer them in this order.

First: what are your data rules? This one can override everything else. If your data cannot legally or contractually leave your control — regulated industry, public sector, sovereignty requirement (Chapter 16, Chapter 23) — then the hosted public API may be off the table no matter what the math says, and your real choice is between private deployment and full self-hosting. Settle this first, because it can shrink the menu before cost ever enters the conversation. This is, in the end, the whole reason Mistral's spectrum exists.

Second: what is your volume, and how steady is it? If your traffic is low, spiky, uncertain, or just starting, the per-token API almost always wins, because you pay only for what you use and carry no fixed burden through the quiet stretches. If your traffic is high, steady, and predictable, the fixed-cost economics of self-hosting start to win, because that big fixed compute cost gets divided across enormous usage until the per-request cost falls below what the meter would have charged. Somewhere between those extremes is a crossover point where the two cost models meet — and finding *your* crossover, with current rates and your real volume, is the single most useful sum you can do before committing.

The crossover sum is worth walking through once, in shape, so you can run your own. Take your real monthly volume — say it works out to some number of millions of tokens a month at your average request size — and multiply by today's verified API rates to get a monthly hosted bill. Then estimate the fully loaded monthly cost of a self-hosted deployment that could carry that volume: the hardware amortized or the cloud compute rented, plus a fair share of the salary of whoever keeps it running, plus power and overhead. Compare the two. Below some volume the hosted bill is plainly smaller and self-hosting is a vanity; above some volume the fixed self-hosted cost is plainly smaller and the meter is a leak. The crossover is where they meet, and the honest version of the sum includes the operating salary, which the optimistic version always forgets — which is exactly why the third question exists.

Third: what can your team actually operate? Self-hosting a model is not free even when the weights are. It demands hardware, machine-learning operations skill, and the ongoing care of keeping a production system healthy (Chapter 15). A team without that capacity can easily spend more in salary, downtime, and headaches than the hosted API ever would have charged, and the "cheaper" self-hosted bill turns out to be an illusion that ignored the cost of running it. Be honest about what you can carry. The right station is the one that fits not just your data and your volume but your actual operational muscle.

There is a frequent failure mode worth flagging here, because I have watched it cost real money: the team that does the token math, sees self-hosting come out cheaper at their volume, and moves — having counted only the compute and none of the operations. Six months later the "cheaper" deployment has consumed an engineer's attention, suffered a couple of outages that sent traffic nowhere, and lagged behind on updates, and the all-in cost has quietly exceeded the API bill they fled. The math was not wrong; it was incomplete. The private-cloud middle station exists precisely for teams who need more control than the public API but cannot honestly carry the full operational weight of self-hosting — it buys back much of the operational burden while keeping the data-handling guarantees, and for a lot of regulated organizations it is the mature answer that neither extreme provides.

The Discipline to Carry Out of This Part

A handful of durable practices keep deployment cost from becoming a surprise, and they outlast every number on the page.

Prototype on the hosted API regardless of where you will end up. It is the fastest way to prove the workload and measure your real volume, and you cannot intelligently choose a station until you know your actual traffic and your actual per-request behavior. Optimize after you have measured, not before.

Pull the model-choice lever first. As Chapter 19 argued, which model answers which request is the largest cost lever on the per-token meter by a wide margin — routing routine work to cheaper, smaller models and reserving the flagship for the hard parts changes the bill more than almost any deployment decision.

Cut the context before you cut anything else. As the worked numbers above showed, the input you send on every call is frequently the dominant cost, not the output you read. Trim retrieved passages to what is actually relevant, drop instruction text the model has already internalized, and do not pass a long conversation history when a short summary would do. This is usually the single largest saving available on the hosted meter and it costs you nothing but attention.

Run the crossover sum with current numbers, not these. The reference rates in this chapter are a magnitude, not a quote. Before you commit to self-hosting on a cost argument, do the arithmetic with today's verified API rates against your real volume and your real, fully loaded operating costs — the figures here will have drifted, and the crossover moves with them.

Cap and watch the meter. On the hosted API, limit reply lengths where long output is not needed, trim the context you send, and set whatever spend alerts the console offers, so a runaway loop trips a wire instead of a bill.

The honest summary is the one this Part has circled from the start. Mistral gives you something rare: not one way to run a capable model but a spectrum, from a per-token API you start in minutes to open weights you host yourself and pay for in compute. The structure of that spectrum — and the two genuinely different cost models at its ends — is stable enough to plan a business around. Every rate attached to it is not. Verify the numbers, locate your station by data rules first and volume second and operational capacity third, and let Mistral's own pricing — never this page — tell you what a token costs today.

Part VII: Mistral for Work

Audience: business and technical decision-makers weighing Mistral for real, repeated use.

"Capability is no longer the hard part. Permission to deploy it on your own data is — and that is the problem Mistral's openness was built to solve."

— A working principle

Chapter 22: Business Use Cases

A logistics company I advised had a problem that no demo ever shows you. They had twenty years of shipping records, contracts, and internal procedures sitting in a document store, and they wanted an assistant that could answer questions against all of it — "what's our standard liability clause for refrigerated freight," "which carriers have we had claims disputes with," that sort of thing. Every off-the-shelf assistant could do it technically. None of them could do it on their terms, because the data included customer contracts they were contractually forbidden to send to a third party, and a clause in those very contracts said so. The capability was a solved problem. The permission was the whole obstacle. They ended up running a Mistral model inside their own network, pointed at their own documents, with nothing leaving the building — and the thing they could not buy from anyone else was not intelligence but the right to use it on their data.

That is the shape of almost every real Mistral business case I have seen. The interesting question is rarely "can the model do this," because at the frontier most assistants can. The interesting question is "can this organization actually deploy it, given what its data is and what its rules are" — and that is where the openness-and-sovereignty combination from Chapter 1 stops being philosophy and starts being the reason a project ships at all. This chapter walks through the concrete patterns where that combination drives real adoption.

Internal Assistants on Private Data

This is the single most common serious deployment, and the logistics story is the template. An organization has a large body of internal knowledge — policies, contracts, technical documentation, support histories, research — and wants an assistant that can answer questions and draft work grounded in *that* material rather than the open internet. The technical pattern is well understood; you connect a model to your documents so it can retrieve and reason over them.

What makes Mistral the choice here is rarely the answer quality in isolation. It is that the private data can stay private. For a law firm, a hospital, a bank, or a defense contractor, "the assistant reads our confidential files" and "our confidential files are sent to a third party's cloud" are not separable in an off-the-shelf product — accepting the first means accepting the second. With Mistral you can run the model where the data already lives, so the assistant reads the files without the files ever leaving your control. The capability becomes reachable for exactly the institutions that were locked out, which is the through-line of this whole book.

Picture how the logistics deployment actually behaves once it is live, because the value is in the ordinary moments. A claims handler is staring at a dispute over a spoiled refrigerated shipment and asks the internal assistant what the company's standard liability terms are for that freight class and whether this particular carrier has a history of similar disputes. The assistant retrieves the relevant contract clauses and the prior claims records from the company's own store, and answers with citations the handler can open and verify. Twenty minutes of digging through a document management system collapses into a question and a checkable answer — and not one byte of those customer contracts left the building to make it happen. The same instinct that would have killed the project under an off-the-shelf product is the instinct the deployment was designed to satisfy.

A practical note worth carrying: an internal assistant is only as good as the documents behind it and the retrieval that feeds it. Teams that succeed here invest as much in organizing and chunking their knowledge as in the model. The model is the easy part now. The failure mode I see most often is a team that bolts a brilliant model onto a messy, duplicated, half-outdated document store and is then baffled when the assistant confidently cites a superseded policy — because it faithfully retrieved exactly the stale document the store still contained. Garbage in, confident garbage out. The unglamorous work of curating the knowledge base is where these projects are actually won or lost.

Document Processing at Scale

The second pattern is less glamorous and quietly enormous: turning unstructured documents into structured, usable information. Invoices, contracts, claims forms, regulatory filings, medical records, research papers — organizations drown in documents that humans currently read one at a time. A capable model can extract the fields, summarize the content, classify the type, flag the anomalies, and route the result.

Mistral fits this case for two reasons beyond raw capability. First, document processing is often *high volume*, and high volume is where the cost story

(Chapter 21) matters most — paying per token to a closed API for millions of documents adds up in a way that self-hosting on your own hardware does not, and at scale the economics can flip decisively. Second, the documents in question are frequently the sensitive kind — financial, medical, legal — which drags us straight back to data control. A bank processing loan applications cannot casually stream applicant financials to a foreign black box. Running the extraction on its own infrastructure, it can.

Consider an insurer processing motor claims as a concrete shape for the pattern. Each claim arrives as a bundle — a submitted form, photographs of damage, a repair estimate, sometimes a police report — and a human used to open every bundle, key the relevant fields into the claims system, classify the claim type, and flag anything odd for closer review. A document-processing deployment reads each bundle, extracts the structured fields the downstream system needs, classifies the claim, and surfaces the anomalies — an estimate wildly out of line with the damage described, a date that does not match the report — for a human to judge. The humans stop doing the keying and the sorting and spend their attention on the exceptions, which is where their judgment was always worth most. The volume is enormous and the data is exactly the kind that cannot be casually exported, which is why this pattern and Mistral's deployment story fit each other so naturally.

The honest caveat: document processing demands accuracy you can verify, not just plausible output. The successful deployments I have seen pair the model with validation — confidence thresholds, human review of low-confidence extractions, audit logs. The model accelerates the work; it does not get to be unaccountable for it. The insurer above does not let a low-confidence extraction flow silently into a payout; below a confidence threshold the claim routes to a human, and every automated decision is logged so an auditor can reconstruct what happened and why. The discipline is what makes the speed safe.

Customer Support

Support is the use case every vendor leads with, and for once the hype is mostly earned — assistants are genuinely good at drafting responses, summarizing long ticket histories, suggesting answers from a knowledge base, and handling routine queries end to end. The capability is real and broadly available across all the major assistants.

Where Mistral earns its place is, again, at the edges that matter to specific buyers. Support conversations contain customer personal data, which in Europe means GDPR obligations the next chapter is devoted to. A European company can deploy Mistral with EU data residency and keep customer conversations inside the jurisdiction its regulators care about — a structurally cleaner story than routing everything through a US-based provider. And because the model is open, a support team can fine-tune it on its own past tickets and house style (Chapter 20), producing an assistant that sounds like the company rather than a generic helper. Lock-in matters here too: support is mission-critical once it is live, and building it on weights you control rather than an API you rent means no single vendor can change your terms out from under a system your customers depend on.

It is worth being specific about the spectrum of support deployments, because "AI for support" hides very different levels of risk. At the cautious end, the model never speaks to a customer at all — it sits beside the human agent, summarizing the ticket history so the agent does not have to read forty back-and-forth messages, and drafting a suggested reply the agent edits and sends. The human stays in the loop on every word that reaches the customer, and the deployment is low-risk because nothing is automated past the point of human judgment. Further along, the model handles the genuinely routine queries end to end — "where is my order," "how do I reset my password" — and escalates anything outside its confident range to a person. The further you push toward full automation, the more the validation discipline from the document-processing pattern reappears: confidence thresholds, clean escalation paths, and logging, because a support assistant that confidently invents a refund policy is worse than no assistant at all. The European residency and the fine-tuned house voice are what make Mistral attractive across that whole spectrum; the discipline about where humans stay in the loop is what makes any point on it safe.

Coding and Software Development

Mistral's coding strength deserves its own line in any business case, because software teams are among its most natural adopters. Through Vibe's Code mode (Chapter 6) and the models behind it, Mistral handles the full range of development work — generating code, explaining unfamiliar systems, writing tests, reviewing changes, debugging.

The business reason a development organization reaches for Mistral specifically tends to be one of three. **Source code is intellectual property**, and many companies are deeply uncomfortable streaming their proprietary codebase to an external service; running an open model internally keeps the code in-house. **Cost predictability** matters for teams where dozens of developers hit the assistant constantly all day, and self-hosting converts an unpredictable per-token bill into a fixed infrastructure cost. And **customization** — fine-tuning on a company's own codebase, conventions, and internal libraries — produces an assistant fluent in your code rather than code in general. None of these is about the model being smarter than its rivals; each is about control over a sensitive, high-volume, high-value workflow.

The intellectual-property concern is more concrete than it sounds in the abstract. A defense contractor or a company whose entire value sits in a proprietary trading system or a patented algorithm cannot have that code traveling to an external service on every autocomplete and every "explain this function," no matter how trustworthy the contract — the security review simply will not pass. For these organizations the choice is not "which coding assistant is best" but "which coding assistant can we use at all without our crown jewels leaving the building," and an open model they run on internal infrastructure is frequently the only candidate that survives the review. The customization angle compounds it: a model fine-tuned on the company's own conventions and internal libraries stops suggesting the generic pattern and starts suggesting *their* pattern — using the internal logging wrapper instead of the standard one, following the house error-handling style — which is exactly the fluency that turns a capable general coder into a teammate who already knows how this codebase does things.

Regulated and Sovereign Workflows

The last pattern is the one Mistral was arguably built for, and it is less a single use case than a category that cuts across all the others: workflows where the *deployment* is the requirement, not a preference. A government ministry that may not place citizen data on foreign infrastructure. A hospital network bound by health-data rules. A defense supplier under sovereignty mandates. A financial institution under regulatory data-residency obligations.

For this category the calculus is different from everything above. These organizations are not weighing Mistral against ChatGPT on features and picking the nicer one. They are filtering the entire market down to the options that clear a legal bar, and frequently the open, self-hostable, EU-resident option is the only one left standing. Public-sector sovereign deployments — an institution running Mistral entirely within its own controlled environment, beholden to no outside provider — are the purest expression of the company's reason to exist. For a great many of these buyers Mistral is not the best choice among several; it is the only choice that is permitted at all.

What this category really demonstrates is that the four patterns above are not separate buyers — they are the same buyer with a constraint switched on. The records office that wants an internal assistant on its citizen correspondence, the ministry processing benefit applications at volume, the public-health line drafting replies to citizens, the agency building internal tooling on sensitive code: each of those is one of the earlier patterns, run by an organization that legally cannot reach for the off-the-shelf version. The sovereignty requirement does not invent a new use case so much as it removes every option except the deployable one, and that is precisely the situation Mistral's spectrum was designed to serve. When you see a sovereign deployment, you are usually watching one of the four ordinary patterns being run by someone who had no other door.

The Pattern Behind the Patterns

Step back from the five cases and the same engine drives all of them. In every one, the raw capability is something multiple assistants could provide — internal Q&A, extraction, support drafting, coding, all of it is broadly available at the frontier now. What is *not* broadly available is the ability to do the work on data you cannot export, at a cost you can predict and contain, customized to your domain, without surrendering control to a single vendor who sets the terms. That bundle — openness, sovereignty, cost control, no lock-in — is what turns a capable model into a deployable one for the organizations in this chapter.

So when you build a Mistral business case, resist the temptation to argue it on benchmark capability, because that is the ground where the case is weakest and the US flagships often lead. Argue it on the bundle. The question that wins budget is not "is this the smartest model" but "is this the model we are actually allowed to run, on our data, at our scale, on our terms" — and for a large and growing set of serious organizations, the honest answer points to Mistral.

One caution to carry forward, because it is the spine of the next chapter: none of these deployments is compliant just because it runs on Mistral. The tool makes compliance *achievable* — it does not make it automatic. A sovereign deployment configured carelessly is still a liability. What openness buys you is the ability to do the work right; doing it right is still work. That distinction is exactly where the European compliance story begins.

Chapter 23: The European Compliance Story

Here is the sentence that ought to govern this entire chapter: the data staying in the building is necessary, and it is not sufficient. Mistral's openness and EU residency hand a European organization the strongest raw material available for an AI compliance story — and they still do not, by themselves, make anything compliant. Compliance is something you *do*, continuously, with process and documentation and accountable humans. The tool determines whether doing it is even possible. It does not do it for you. Hold both halves of that at once and the rest of this chapter falls into place.

Why Compliance Is Mistral's Strongest Ground

Everywhere else in this book I have been careful to say Mistral is not always the most capable model and that the US flagships often lead on the raw capability ceiling. Here, on European compliance, the ranking genuinely flips — this is the ground where Mistral's structural advantages are decisive rather than marginal. The reason is not marketing. It is architecture. A company that publishes its weights and operates EU-based infrastructure can offer a European buyer things a US-based, API-only provider structurally cannot, no matter how good that provider's intentions or contracts are. The next sections are the specifics.

The word "structurally" is carrying real weight there, and it is worth slowing down on, because it is the difference between an advantage that can be competed away and one that cannot. A US-based provider can write you an excellent contract, open European data centers, promise zero retention, and mean every word of it — and still a European regulator can observe that the provider remains a US company subject to US legal processes that may, in principle, reach data it holds anywhere in the world. No contract closes that gap, because the gap is not contractual; it is jurisdictional. Mistral's advantage on this ground is of the kind that cannot be matched by a better contract or a bigger data-center budget, because it lives in where the company is incorporated and where the model can run, not in the terms it offers. That is what makes compliance the one chapter where I will say plainly that Mistral leads.

GDPR: Control Is the Currency

The General Data Protection Regulation has governed personal data in the EU since 2018, and it is the baseline every European AI deployment must satisfy. Its logic, stripped down, is that personal data belongs to the person it describes, and any organization processing it carries real obligations — a lawful basis for doing so, limits on what it collects and keeps, security over what it holds, and genuine accountability for all of it.

AI complicates this in a specific way: the moment you send personal data to a model, that transmission is itself a processing event with consequences. Where did the data go? Who can see it? Is it used to train the provider's future models? Under whose jurisdiction does it now sit? For a closed, API-only assistant, the answers are the provider's to give and yours to accept — and for a US-based provider, "under another country's legal reach" is an answer European regulators have grown steadily less comfortable with.

Mistral's openness changes the available answers. Run the model yourself and the personal data never leaves your environment, which collapses a whole class of GDPR questions into "we processed it on our own controlled systems" — by far the cleanest posture available. Use Mistral's EU infrastructure instead and you get data residency within the jurisdiction, processed under EU law, by a provider whose entire identity is built on that promise. Either way, the currency GDPR cares about is *control*, and control is precisely what openness buys.

It is worth naming the specific GDPR concepts where this control pays off, because "cleaner posture" is vague and the obligations are not. Under GDPR you are the *data controller* — the party that decides why and how personal data is processed — and any service that processes it on your behalf is a *processor*, a relationship that normally requires its own contract and its own assurances. Self-host the model and a large part of that processor relationship simply dissolves, because there is no third party in the loop; you are processing on your own systems and the chain of accountability stops with you. The training question gets cleaner too: a recurring GDPR worry with closed assistants is whether the personal data you send becomes training fodder for the provider's future models, a secondary use you did not authorize. On a self-hosted open model there is no provider receiving the data to train on, and on Mistral's EU infrastructure the use of your data is governed by terms built around the European posture — **the**

specifics of which you should confirm in the current terms rather than assume. And the *transfer* question — GDPR's restrictions on moving personal data outside the EU — barely arises when the data never crosses the border in the first place.

The honest line: GDPR compliance is never just "where does the data sit." It is also lawful basis, data-minimization, retention limits, subject-access rights, breach procedures, and your records proving all of it. Mistral hands you the residency and control piece, which is the piece that is genuinely hard to retrofit. The rest is your program to run.

The EU AI Act: A Phased Obligation, Not a Single Switch

The EU AI Act is the world's first comprehensive AI law, and any European organization deploying AI now lives under it. Two features matter most for planning. First, it is **risk-tiered**: obligations scale with how risky a system's use is, from minimal-risk uses with light requirements up through high-risk uses — and a handful of prohibited ones — carrying heavy duties around transparency, documentation, human oversight, and data governance. The same model can land in different tiers depending on what you point it at; an assistant drafting marketing copy and an assistant screening job applicants are not treated alike.

That tiering point deserves an example, because it is the part organizations most often get wrong about themselves. The risk tier is a property of the *use*, not the model. Take a single Mistral deployment inside one company: pointed at drafting internal marketing copy, it is a low-risk use carrying light obligations; pointed at screening job applications or scoring creditworthiness, the same underlying model is now doing something the Act treats as high-risk, with heavy duties around human oversight, documentation, and the quality of the data it relies on. This is why "is our AI use high-risk?" has no single answer for an organization — it has one answer per use. The practical consequence is that you classify each workflow on its own, and a company can quite reasonably run a low-ceremony deployment for one purpose and a heavily governed one for another, on the very same model. Getting this classification right is the first real act of AI Act compliance, and it is one no tool can do for you.

Second — and this is the part that trips up planning — the Act is **phased**. Its obligations do not all arrive at once; they switch on in stages across a multi-year timeline, with different categories of requirement becoming enforceable at different points, and major provisions escalating through 2026. The dates and exact scope are the most perishable facts in this chapter — the phasing has been the subject of active debate and adjustment, so treat any specific deadline I might name as something to verify against the current official text rather than a fixed point. What is durable is the shape: this is a staged rollout, the obligations get heavier over time, and "we'll deal with it when it's fully in force" is not a strategy, because parts are in force now and more arrive on a schedule you can plan against. To take one concrete anchor that the book has used elsewhere: the Act's provisions on prohibited practices and on AI literacy under Article 4 became applicable on 2 February 2025, while broader obligations phase in later and major enforcement escalates through 2 August 2026 — **dates worth re-confirming against the official timeline, since the phasing has moved.**

How does Mistral's openness help with the Act specifically? Several of the Act's harder demands — transparency about how a system works, data governance, human oversight, documentation for high-risk uses — are easier to satisfy when you can actually *inspect and control the model.* You can audit an open-weight model in ways a black box does not permit; you can document its behavior because you can examine it; you can govern the data flowing through it because the flow happens on your own infrastructure. Openness does not exempt you from a single obligation. It makes the obligations you do have meaningfully more tractable.

Data Residency, Concretely

Strip away the acronyms and data residency is a simple, stubborn requirement: certain data must physically remain within certain borders, and must not fall under foreign legal reach. European public bodies face this explicitly; regulated industries face it through sector rules; a growing number of private companies adopt it as policy because their customers or their own risk teams demand it.

This is where a US-based provider hits a structural wall that no contract fully closes. Even a US company operating European data centers remains a US company subject to US legal processes that can, in principle, reach data it holds anywhere — and European buyers and regulators know it. Mistral's answer operates at a different layer: run the model on infrastructure *you* control and the residency question is yours to answer absolutely, or use Mistral as a European provider offering EU residency under EU law without the foreign-reach overhang.

For an organization whose requirement is "this data does not leave the jurisdiction and does not fall under another country's authority, full stop," that structural difference is the whole decision.

How Openness and Self-Hosting Earn Their Keep

Pulling the threads together, the mechanism by which Mistral's defining traits help compliance is consistent across GDPR, the AI Act, and residency rules alike. **Self-hosting keeps data in your control** — the cleanest answer to nearly every "where did it go and who can reach it" question a regulator can ask. **Open weights let you audit and document the model**, which serves the transparency and governance demands a black box cannot. **EU residency**, whether self-hosted or on Mistral's infrastructure, keeps you inside the jurisdiction your regulators care about. **No vendor lock-in** means your compliance posture does not sit at the mercy of a single provider's unilateral changes to terms or location.

The lock-in point is more load-bearing for compliance than it first appears, and it is easy to miss. When your AI compliance argument depends on where a closed provider runs its servers and what its terms say, that argument can be undone by a decision you do not control: the provider moves a workload, revises its retention policy, changes which subprocessors it uses, or alters where data is handled — and a compliance posture you documented carefully last year quietly stops being true, often without your noticing until an audit finds it. An open-weight model you run yourself cannot have its data-handling rewritten by someone else's product decision, because there is no one else in the loop to rewrite it. The compliance posture you established is the posture that persists, which is a kind of stability that matters enormously when you have to attest, year after year, that the system still does what your documentation says it does.

None of these is an abstraction. Each maps to a specific obligation that is genuinely difficult to meet with a closed, foreign, API-only assistant and genuinely tractable with an open, sovereign, EU-resident one. This is the compliance officer's necessary condition — the raw material without which the rest cannot be built.

Compliance Is a Process, Not a Product

Now the other half, and the reason this chapter opened on necessary-but-not-sufficient. Everything above describes what Mistral *makes possible*. It does not describe compliance, because compliance is not a property a tool confers. It is an ongoing organizational practice: a lawful basis documented, a data-protection impact assessment performed, a risk tier determined and recorded, human oversight designed into the workflow, retention and access controls configured, audit trails kept, incidents handled, the whole apparatus reviewed as the law evolves.

You can run Mistral fully self-hosted inside an EU data center and still be out of compliance — by processing personal data without lawful basis, by skipping the impact assessment a high-risk use requires, by configuring access carelessly, by keeping no records. The tool removed the hardest structural obstacles; it did not perform the program. And the inverse warning matters just as much: do not let the strength of Mistral's compliance story breed complacency. "We use the European, open, sovereign option" is the *start* of a compliance argument, the part that makes the rest achievable. It is not the argument itself.

So the right way to hold this chapter is as a two-part claim, both parts true at once. Mistral gives European organizations the strongest available foundation for AI compliance — that is genuine, and it is the clearest case in the entire book for choosing Mistral over a US flagship. And a foundation is not a finished building. The work of compliance — the assessments, the documentation, the oversight, the accountable humans — remains yours to do, on top of a foundation that, with Mistral, you are finally allowed to build at all.

That same distinction between what the tool provides and what the buyer must still supply is the kind of honest accounting that decides the larger question too: put Mistral squarely beside the US flagships and ask, without flinching, where each one actually wins.

Chapter 24: Where Mistral Fits vs. the US Flagships

The most useful sentence I ever heard about this question came from a CTO who had spent a month evaluating Mistral against ChatGPT, Claude, and Gemini for his company. “We stopped asking which one is best,” he said, “the day we realized we were going to use two of them.” That is the conclusion this chapter is built to earn honestly, and I want to be plain about my obligation here: this book makes Mistral’s case, but it is a guide, not an advertisement. A guide that told you Mistral was the right tool for every job would be doing you a disservice and you would catch it eventually anyway. So this is the even-handed accounting — where Mistral genuinely leads, where the US flagships genuinely lead, and how a sensible organization actually decides. No invented benchmarks; the comparison that matters is structural and durable, not a leaderboard that turns over every quarter.

Start With What’s Shared

Before the differences, the unglamorous truth that frames everything else: at the frontier, these systems are more alike than the marketing on any side admits. Mistral’s top models, ChatGPT, Claude, and Gemini are all capable assistants that handle the bulk of ordinary professional work — writing, summarizing, analysis, coding, answering questions, reasoning through a problem — at a quality that would have seemed impossible a few years ago. On a typical task, run blind, most people could not reliably tell which one produced the result. If you already use a US flagship and are happy with it, Mistral will not feel like a different species. It will feel like a strong peer with a sharply different deployment story.

This matters for the comparison in a way that is easy to skip past: because the systems are so close on ordinary work, capability is rarely the deciding axis for serious deployment decisions. When four assistants all do the writing, the summarizing, and the routine coding well enough that a blind test cannot separate them, the choice among them stops being “which is smartest” — there is no meaningful daylight there for most work — and becomes “which fits the constraints around the work.” That shift is the entire reason the rest of this chapter is organized around *where* each option leads rather than *how much*. A comparison built on capability scores would be measuring the axis on which the options are most alike and ignoring the axes on which they actually differ.

That framing is the whole key. The question is almost never "which single assistant is smartest at everything" — none is — but "which one fits *this* requirement," and once you ask it that way, the lines come into focus.

Where Mistral Genuinely Leads

Openness — this is the foundational one. Mistral publishes its weights under permissive licenses; the US flagships do not. That single fact cascades into everything that follows in this section. You can run a Mistral model yourself, inspect how it works, fine-tune it deeply on your own data, and deploy it without anyone's permission. With a closed flagship you rent access to a black box on the provider's terms. For organizations that need to control, audit, or own their AI rather than rent it, this is not a feature comparison — it is a different category of relationship, and it is the one place where no US flagship offers an equivalent at all.

It is worth being precise about why this is a category difference and not a degree difference, because that is exactly the kind of claim that sounds like advocacy until you look at it closely. Every other advantage on this list — better residency terms, lower cost, more favorable contracts — is in principle something a closed provider could match if it chose to; they are points on shared axes. Openness is not on a shared axis. A closed provider cannot offer you a slightly more open black box; the weights are either published or they are not, and the US flagships have chosen not to publish theirs. So when an organization's requirement is genuinely "we must be able to inspect, modify, and run this model ourselves," there is no negotiation that gets a closed flagship there. It is the one Mistral advantage that is structurally unmatchable rather than merely currently better, which is why I keep returning to it as the foundation under all the rest.

Data sovereignty and EU residency. As Chapter 23 laid out, an open, EU-resident, self-hostable model gives European organizations a compliance and residency posture a US-based provider structurally cannot match — not for lack of effort, but because a US company remains under US legal reach wherever its servers sit. For buyers whose hard requirement is that data stays in the jurisdiction and outside foreign authority, Mistral often leads not by a margin but by being the only option that clears the bar.

Cost control. Closed flagships meter access per token on the provider's pricing. Self-hosting an open model converts that into a fixed infrastructure cost you own — and at high volume, the economics can flip decisively in Mistral's favor (Chapter 21). Just as important, you are insulated from a provider raising prices on a system you now depend on. Control over cost, not merely lower cost, is the durable advantage.

Freedom from lock-in. Build on weights you control and no single vendor can change your terms, deprecate your model, or alter its behavior out from under you. Build on a closed API and you have a dependency you do not govern. For mission-critical systems meant to run for years, that independence is worth real money and real peace of mind.

There is a subtle version of the lock-in advantage that the headline understates. With a closed flagship, the provider can deprecate the exact model your system was built and tested against and retire it on its own schedule — and teams that built carefully around a specific model's behavior have, more than once, been forced into an unplanned migration when that model was sunset, re-validating everything against a successor they did not choose and did not time. An open-weight model you have downloaded does not get retired out from under you; the weights you hold keep working regardless of what the company does next. For a system meant to run for years under stable, attestable behavior — which is most regulated systems — "the model cannot be taken away from us" is a different and stronger guarantee than "the model is cheaper," and it is one the closed providers structurally cannot offer.

Where the US Flagships Genuinely Lead

A comparison you can trust has to be just as plain about the other direction.

The raw capability ceiling. The largest US labs run more compute and larger research operations, and on the most demanding tasks — the hardest reasoning, the most sophisticated work at the absolute edge of what is possible — their top models often hold a lead. Mistral is frontier-*class* and genuinely competitive; its argument, as Chapter 1 said, is "highly capable *and* open *and* sovereign," not "always the smartest." When you need the highest capability ceiling available and the deployment constraints are not binding, a US flagship is frequently the stronger pick, and pretending otherwise would not help you.

The honest nuance here is that the capability lead, where it exists, lives at the *ceiling* rather than across the whole range — and most work does not run at the ceiling. The flagships tend to pull ahead on the genuinely hard edge cases: the most tangled multi-step reasoning, the most demanding novel problems, the tasks where a small capability difference actually changes the outcome. On the broad middle of ordinary professional work — the writing, the summarizing, the routine analysis and coding that fills most people's days — the gap is small to invisible, as the opening section said. So "the flagships lead on capability" is true and worth saying, but it is a claim about the top few percent of difficulty, not about your typical Tuesday. The practical question is whether *your* hardest tasks actually live at that ceiling, because if they do not, the capability lead is real and largely irrelevant to you.

Ecosystem breadth. The US incumbents sit inside vast, mature ecosystems — deep developer tooling, enormous communities of shared techniques, third-party integrations, and in Google's case a tight weave into Workspace, Android, and Search. "It's already inside the tools we use all day" is a real convenience the incumbents hold and Mistral is still building toward.

Multimodal range. The US flagships have generally pushed the breadth of multimodal capability — across images, audio, video, and richer real-time interaction — further and faster. Mistral is capable across modalities, but if your work centers on the widest, most polished multimodal range, the flagships often lead here today. This is a fast-moving front, so treat it as a current snapshot rather than a fixed law.

Consumer polish. For the individual just wanting the smoothest possible out-of-the-box assistant experience — the most refined app, the most frictionless features — the US incumbents have poured enormous effort into consumer surfaces. Mistral's deepest advantages live at the enterprise and sovereignty layer, exactly as Chapter 1 warned; for the casual individual user, the sovereignty story is abstract and the polish gap is the thing they will actually notice.

It is worth noticing what kind of advantages these four are, because it explains why they sit on the opposite side of the ledger from Mistral's. The flagship strengths — capability ceiling, ecosystem, multimodal range, polish — are all advantages of *scale and incumbency*: they flow from more compute, more engineers, more years in market, more users to learn from. They are real, and they are also the kind of advantage that a well-funded competitor narrows over time, which is why each one comes with a "today" or a "for now" attached. Mistral's advantages, by contrast, are advantages of *structure* — openness, residency, control — which is why I refuse to hedge them with "for now." The flagships lead on the things money and time buy; Mistral leads on the things incorporation and architecture decide. Knowing which kind of advantage you are weighing tells you how durable it is likely to be.

How to Actually Choose

Forget "which is best" and ask "which for this." A few rules of thumb, all starting points rather than laws.

Choose Mistral when control, openness, or compliance is the binding constraint. If you must run on your own data without exporting it, must satisfy EU residency, must audit or own the model, must contain cost at high volume, or must avoid single-vendor dependence — Mistral is the natural home, and frequently the only option that qualifies. These are the cases where its lead is decisive rather than marginal.

Choose a US flagship when you need the absolute capability ceiling or the deepest ecosystem and your deployment constraints leave you free to. For the hardest tasks at the edge of the possible, the widest multimodal range, or work that lives inside an incumbent's ecosystem, the flagship is often the stronger tool, and there is no shame in saying so in a Mistral book.

Run both — this is the quietly correct answer for a lot of serious organizations. The CTO who opened this chapter landed exactly here. Use Mistral for the sovereign, regulated, cost-sensitive, control-critical workloads it is built for, and reach for a US flagship when a task genuinely needs the capability ceiling and carries no constraint that forbids it. The assistants are cheap relative to the value of getting the work right, and matching each workload to the tool that fits it beats forcing everything through one. Splitting along the control axis — sensitive and regulated work on Mistral, unconstrained ceiling-pushing work on a flagship — is a coherent architecture, not a compromise.

The "run both" answer deserves a word of practical caution, because it is correct in principle and easy to botch in practice. The split that works is a *deliberate* one along a clear axis — typically the control axis, with regulated and sensitive work routed to a model you can run sovereignly and unconstrained ceiling-pushing work routed to a flagship — written down as policy so people know which tool to reach for and why. The split that fails is the accidental one, where different teams quietly adopt different assistants for no articulated reason and the organization wakes up with confidential data flowing to a closed provider because someone defaulted to the convenient app without checking the constraint. Running both multiplies your governance surface: two sets of terms, two data-handling postures, two things to keep an eye on. That is a perfectly reasonable cost to pay for using each tool where it fits, but only if the routing rule is intentional and known, not emergent and discovered later by an auditor.

The Comparison That Matters Is Yours

Any specific "Mistral beats X on benchmark Y" claim I put here would be stale within a quarter and would teach you nothing durable, so I have deliberately anchored none of this to scores. The comparison that actually matters is not on a leaderboard; it is the one you run on your own real requirements. Take your genuine constraints — your data rules, your residency obligations, your volume, your capability needs, your tolerance for lock-in — and weigh the options against *those*. You will learn more from that exercise than from any benchmark table, and you will quickly feel where Mistral's openness earns its place and where a flagship's capability ceiling still wins.

What you should carry out of this chapter is the lay of the land. Mistral leads, often decisively, wherever control, sovereignty, compliance, cost, and independence are the deciding factors. The US flagships lead wherever the raw capability ceiling, ecosystem depth, multimodal range, or consumer polish is what you are buying. Neither wins everywhere, the honest answer for many organizations is to run both, and the only comparison that settles it is the one against your own requirements — which is the kind of clear-eyed judgment the final part of this book is entirely about.

Part VIII: Judgment and Limits

Audience: everyone — and required reading before you commit to Mistral for anything that matters.

> *"The same control that makes Mistral usable where others aren't is also responsibility it hands you. The freedom and the duty arrive together."*
>
> — A working principle

Chapter 25: What Mistral Gets Wrong

A book that only told you what Mistral does well would be a brochure, not a guide. The truth is that Mistral, like every AI company and every large language model, gets things wrong — and it gets two distinct *kinds* of things wrong, which is why this chapter comes in two halves. There are the failures Mistral's models share with every other model on the market, the ordinary fallibilities of the technology. And there are the failures that are specifically Mistral's — the places where being a smaller, open, European challenger to the American giants comes with real costs, not just virtues. An honest guide owes you both. Knowing them is not pessimism; it is what lets you point your skepticism where it pays off and avoid the situations where the tool is weakest.

The Model Failures Mistral Shares With Everyone

Start with the failures that are not unique to Mistral at all, because they shape how you should read anything its models produce.

Confident wrongness. Mistral's models, like all of them, do not reliably signal their own uncertainty. A correct answer and a subtly wrong one arrive in the same fluent, assured tone. You cannot use the model's confidence as a gauge of its correctness, because the confidence is roughly constant whether the content is right or wrong. This is the deepest reason your judgment stays in the loop: plausible-looking-but-wrong is the most dangerous output, because it sails through a quick skim and only careful reading catches it.

To see how this bites in practice, picture a finance analyst who asks Vibe to summarize the year-over-year change in a set of figures she pastes in. The model returns a clean paragraph: revenue up eleven percent, margins compressed slightly, one division down. It reads like exactly the summary she would have written herself, which is the problem — the eleven percent is right, but the model has transposed two of the division figures, and the "one division down" is actually the one that grew. Nothing in the tone flags the slip. The summary that is ninety percent correct is more dangerous than one that is obviously garbage, because the obviously garbage one gets thrown out and the plausible one gets pasted into a board deck. The defense is not to distrust everything the model says equally; it is to know which sentence in the paragraph is load-bearing and check that one against the source.

Hallucinated specifics. When a Mistral model does not know something, it can produce a confident, plausible, entirely fictional detail rather than admitting the gap — an invented citation, a fact that sounds right without being right, a function or parameter that does not exist. The defense is the old one: do not repeat a specific you cannot trace to a source, especially citations, quotes, and statistics.

The coding version of this is worth its own mention because it is so common and so easy to catch. Ask a Mistral model to use a library you are unfamiliar with and it may confidently call a method that has the right shape — sensible name, plausible arguments — but does not actually exist in that library's API. The code looks correct, reads correctly, and fails the moment you run it with an "attribute not found" error. This is the friendliest possible form of hallucination, because the runtime catches it for you immediately. The dangerous form is the invented statistic in a document, which has no compiler to flag it and will sit there looking authoritative until a reader who happens to know the real number finds it. The rule that covers both: a specific you cannot trace, you do not ship.

Subtle reasoning and factual errors. Strong on the common case, weaker on the precise — a number slightly off, a step in a chain that does not quite follow, an edge case missed. These survive the "does this look right?" read and only fail where the specific detail matters, so verify the load-bearing specifics rather than spot-checking the prose.

A concrete shape this takes: you give the model a multi-step calculation — a pricing formula with a discount tier, a tax, and a rounding rule, say — and it walks through the steps in order, narrating each one correctly, and then produces a final number that does not match the steps it just showed. The reasoning is right and the arithmetic at the last step is wrong, or the reasoning skips the rounding rule it acknowledged two sentences earlier. Because the *explanation* is sound, a reader nods along and trusts the conclusion. The way to catch this class of error is to treat the final number as a separate claim from the reasoning that produced it and check it independently, rather than letting a convincing chain of steps vouch for its own result.

The limits of training data and scale. Like any model, Mistral's have a knowledge cutoff and can be out of date; they are weakest on genuinely novel problems and on highly specialized, high-stakes domains where a wrong answer is expensive. In those domains the model is a thinking aid and a starting point, never the final authority — and never the holder of the liability, which stays with you.

The cutoff matters more than people expect because the model does not always *know* that it is out of date. Ask about a regulation, a software version, or a product that changed after the training cutoff, and the model will often answer confidently from the world as it knew it, with no flag that the answer might be stale. This is exactly why the web-grounded research mode (Chapter 10) exists and why a sovereignty-conscious user running an older self-hosted model needs to be especially alert: a model you froze on your own hardware eighteen months ago does not get fresher on its own, and its confident answer about "the current rules" is an answer about the rules as of its training, not as of today.

None of that is special to Mistral. It is the cost of using this technology at all, and the defenses are the ones this whole series keeps returning to: skepticism toward confident output, verification of what matters, and human ownership of the high-stakes call.

The Failures That Are Specifically Mistral's

Now the half that an honest guide to *this* company has to add, because choosing Mistral is choosing a particular set of trade-offs, and the costs are as real as the benefits this book has spent twenty-four chapters describing.

It is often not the most capable model in the room. Mistral is a frontier-class lab, but it is a smaller company than the American giants, with less compute and a smaller research org, and on the raw capability frontier — the hardest reasoning, the most demanding coding, the trickiest novel problems — the US flagships frequently lead. If your single requirement is the absolute top of the capability curve and nothing else matters, Mistral is often not the answer, and pretending otherwise would not serve you. Mistral's case is the *combination* of capability with openness and sovereignty (Chapter 24), not a claim to the capability crown.

It helps to be specific about where the gap shows up, because it is not uniform. On everyday writing, summarization, classification, and the broad middle of professional work, you will rarely feel the difference — a capable Mistral model and a capable US flagship produce comparably good results, and the choice comes down to the sovereignty and cost factors this book has emphasized. The gap opens on the hardest tail of the distribution: a competition-grade reasoning puzzle, a sprawling refactor across an unfamiliar codebase, a research synthesis that requires holding many subtle threads at once. If your work lives mostly in the middle of that distribution, the capability ceiling is a number on a leaderboard you will almost never touch. If your work lives in the hard tail every day, it is a real reason to keep a flagship in your toolkit. Knowing which describes you is more useful than knowing which model wins a benchmark.

The ecosystem is thinner. The American flagships sit inside vast ecosystems — integrations, plugins, third-party tools, enormous communities, a deep bench of tutorials and shared techniques. Mistral's surrounding ecosystem is real and growing but smaller. You will find fewer ready-made integrations, fewer answers to your exact problem already written down, fewer people who have walked the path ahead of you. For self-hosting and building, this means more of the work falls on you.

The practical face of this is the moment you hit a wall and go searching for help. With the largest US providers, the odds are good that someone has already posted your exact error message and three people have answered it. With Mistral, especially on a less-trodden path like a particular self-hosting configuration or an unusual fine-tuning setup, you are more likely to find the official documentation and then a thinner trail of community discussion beyond it. The documentation itself is good and improving, but "the docs plus your own experimentation" is a heavier path than "the docs plus a decade of accumulated forum answers." Budget for that. A team adopting Mistral for a build should assume it will occasionally be the first to solve a given problem in public, rather than the thousandth.

Multimodal and feature breadth lag. The biggest US assistants have pushed hard and broad into images, video, voice, and a sprawling feature surface. Mistral's coverage across every modality and feature is generally narrower. If your work depends on a specific capability the flagships have and Mistral has not yet matched, that gap is a real reason to look elsewhere for that task.

The honest move here is to check the current state rather than trust either this book or your assumptions, because this is the fastest-moving axis of all. Mistral has added capabilities steadily — and "Mistral cannot do X" has a way of becoming "Mistral added X last quarter" without much fanfare. The durable advice is not a list of what is missing today, which would be stale by the time you read it, but a habit: before you build a workflow around a specific modality or feature, confirm against the current documentation that Mistral supports it the way you need, rather than discovering the gap halfway through the build. The gap is real where it exists; it is also narrower every year, and pinning your mental model to last year's gaps will steer you wrong in the other direction.

Consumer polish is behind. Vibe is good and improving fast, but the most-used American assistants have had more time, more users, and more resources poured into the consumer experience. You may meet rougher edges — a feature that is less refined, a workflow that is less smooth, documentation that is thinner — than you would on the most polished alternatives.

This is the kind of failure that is easy to overstate and easy to underweight, depending on who you are. For an individual choosing a daily assistant on feel alone, polish matters a lot, and a slightly rougher edge can be the thing that sends them back to the more familiar tool. For a regulated organization choosing a deployment, polish matters far less than whether the data stays where the law requires — a slightly less smooth interface is a rounding error against a compliance requirement that the polished alternative cannot meet at all. The mistake is letting consumer-grade polish judgments drive an enterprise-grade decision, or vice versa. Weigh polish proportionally to how much your actual use depends on it.

The openness is real but not unlimited. This book has celebrated Mistral's open weights, and rightly — but Chapter 14 was honest that "open weight" is not "open source," that licenses vary by model, and that the openness has shifted over time and may shift again. Do not assume every model is equally open, that any given model will stay open forever, or that open weights deliver the full open-source ideal. Read the actual license for the actual model you intend to use, every time, because the company's openness is a strategy, not a constitutional guarantee.

The trap this sets is a planning trap, not a today trap. A team can read "Mistral is open" once, internalize it as a permanent fact, and build a multi-year plan on the assumption that the next model and the one after will land under the same permissive terms as the current one. That assumption is the dangerous part. The right posture is the one Chapter 14 set out: the openness is genuine for the models that have it, the weights you have already downloaded cannot be clawed back, but the terms of *future* releases are a business decision Mistral makes release by release, not a promise you can bank on. Plan around the licenses you can actually read for the models you can actually obtain, and treat any future model's openness as a hope rather than a guarantee until its license is published.

Self-hosting moves the burden to you. The flip side of sovereignty (Part V) is operational responsibility. When you run the weights yourself, you own the infrastructure, the updates, the security, the scaling, and the cost — work the hosted American providers do for you. The control is genuine and, for many, worth it; the burden is also genuine, and underestimating it is a common and expensive mistake.

The expensive version of this mistake looks like a project that ships a self-hosted model in a weekend demo and then discovers, three months in, that nobody owns the operational side. The GPUs cost more than the napkin estimate because the workload runs hotter than expected. A security patch comes out and there is no process to apply it. The model needs to scale for a usage spike and the single box it runs on cannot. None of these are arguments against self-hosting — they are the predictable, manageable costs of it, and the organizations that succeed are the ones that scoped those costs honestly before they committed. The benefit is real and the burden is real, and the failure mode is treating the first as free because you were excited about it and forgetting to budget for the second.

How to Hold Both Halves

The point of this inventory is not to undercut the book — it is to make you a clear-eyed user instead of a disappointed one. Each failure maps to a defense or a decision. The shared-model failures map to the same discipline every assistant requires: verify what matters, trust nothing blindly, keep the high-stakes judgment human. The Mistral-specific failures map to honest expectation-setting: reach for Mistral when its actual strengths — openness, sovereignty, compliance, cost control, freedom from lock-in — are what you need, and reach elsewhere when your binding requirement is the raw capability ceiling, the deepest ecosystem, or the broadest feature set. Many serious organizations run both, using Mistral exactly where its trade-offs win and a US flagship where theirs do.

That is the honest posture, and it is the same one this series brings to every tool: neither the credulous view that Mistral is the one true answer, nor the cynical view that a smaller challenger cannot matter. Mistral is a genuinely capable, genuinely open, genuinely sovereign AI that is genuinely behind on some axes, and using it well means knowing precisely which axis your task lives on. Get that right and you get the benefits this book has described without the disappointment of expecting the things Mistral does not yet do.

The final chapter turns from what Mistral gets wrong by the nature of the technology to what you have to get right by the nature of your responsibility: using a capable, open, self-hostable AI well and safely, especially at work.

Chapter 26: Responsible Use

There is a particular irony to writing a responsible-use chapter for Mistral, and it is worth naming up front. Mistral's whole proposition is *control* — over your data, your deployment, your model. That control is exactly what makes responsible use both easier and more genuinely yours. With a closed American assistant, a lot of the responsibility is taken out of your hands, for better and worse: the provider decides the guardrails, holds the data, sets the terms. With Mistral, especially when you self-host, more of those decisions become yours to make — which means more of the responsibility does too. The freedom and the duty arrive together. This final chapter is about carrying the duty as deliberately as you take the freedom.

The Sovereignty Cuts Both Ways

Self-hosting an open-weight model is the strongest form of data control there is — and it hands you obligations the hosted providers were quietly carrying. When the model runs on your infrastructure, *you* are responsible for securing it, for who can access it, for keeping it patched, for monitoring how it is used, and for the safety behavior that a hosted provider would have enforced at their layer. The data never leaving your walls is a privacy win; it also means no one else is watching for misuse, catching the harmful output, or maintaining the safety scaffolding. Sovereignty is not just a benefit you receive; it is a job you accept. Take it on with your eyes open, and resource it like the real responsibility it is.

Guard What You Feed It — and Know Where It Goes

The first rule everywhere in this series applies here too, with a Mistral-specific twist. Be deliberate about what data goes into which deployment, because Mistral's deployments differ sharply in their data posture.

If you use the **hosted Vibe consumer product**, treat it like any other consumer AI assistant: assume your conversations may be used to improve the service unless the terms say otherwise, and keep genuinely sensitive material out (Chapter 16 covers the privacy specifics). The European data posture is a real advantage, but it is not a license to paste in secrets.

If you use **La Plateforme**, the **enterprise** deployments, or **self-hosting**, the data-handling guarantees are stronger and more contractual — which is precisely why regulated organizations choose them. But strong guarantees still require you to read them and configure for them. The tool *enables* keeping data in-region and under control; it does not do it automatically if you point it at the wrong deployment. Match the sensitivity of the data to the deployment that actually protects it.

The Mistral-specific habit, then, is this: **know which deployment you are using before you decide what to feed it.** The same company offers a consumer chat box and a sovereign self-hosted model; they are not interchangeable for sensitive work.

A Compliant Tool Is Not a Compliant System

Chapter 23 made this point and it bears repeating as a rule of responsible use: choosing a compliant-friendly tool is not the same as being compliant. Mistral's openness, EU residency, and self-hosting make compliance *achievable* in situations where a US-only black box makes it impossible — but GDPR, the EU AI Act, and your sector's regulations impose obligations on *how you operate*, not just which vendor you pick. You still have to do the data-protection assessment, document the processing, govern access, and meet the obligations the regulation places on a deployer of AI. Mistral can be the foundation of a compliant system; it cannot be the whole of one. Treat the tool as necessary-not-sufficient and bring the process.

Verify Before It Becomes a Decision

The confidently wrong problem from Chapter 25 is not academic at work — it is how a bad number reaches a report or a wrong answer reaches a customer. The discipline scales with stakes: a sensible read for low-stakes internal use, independent verification of the load-bearing facts for anything that informs a real decision, goes to a customer, or gets published. "Mistral said so" is not a citation you can stand behind professionally; the source it drew from is. This holds with extra force for fine-tuned and self-hosted models, where you have customized the behavior and cannot lean on a provider's quality bar — the more you have made the model your own, the more its output is your responsibility to check.

Use Open Weights Ethically

Open weights bring a freedom worth using thoughtfully. Because you can fine-tune and deploy the models yourself, with fewer external guardrails than a hosted assistant imposes, you also inherit the responsibility for what your customized model does. Do not use the freedom to strip safety behavior for its own sake; do not deploy a model in a high-stakes setting without the testing and safeguards the setting demands; respect the licenses (they are permissive, not absent — Chapter 14); and remember that "I could run it however I wanted" is a statement about capability, not about what is responsible. The openness is a gift to builders who use it well, and the community of open AI depends on that good faith.

Work Within the Rules — and Help Make Good Ones

Many organizations now have AI-use policies, and a sovereignty-minded organization may have especially specific ones. Know yours: which deployments are sanctioned, what data may go where, what review is required. If your organization chose Mistral specifically for its data-control properties, using a personal consumer account for work data would quietly defeat the entire reason for the choice — use the sanctioned deployment. And if you are one of the people who *writes* the policy, Mistral's spectrum is an asset: you can route the sensitive work to self-hosted or private deployment and the ordinary work to the convenient hosted one, matching the control to the need rather than banning or blessing everything at once.

Where Responsibility Lands

Net it out and the picture is the one this series always arrives at, sharpened by Mistral's particular character. The tool gives you an unusual amount of control, and control is responsibility wearing its working clothes. Use the hosted products with the ordinary care any consumer AI deserves; use the sovereign deployments with the deliberate data-governance they make possible and the regulations require; verify what matters; respect the licenses and the safety the open models depend on; and never let the freedom Mistral hands you become an excuse to be careless with it. The accountability for what you build and decide does not transfer to the model, the company, or the country it comes from. It stays with you — which is, in the end, exactly what sovereignty means.

That word, sovereignty, has run through this whole book as Mistral's defining feature. The short reflection that follows asks what it really means to treat it as a feature at all — and then the reference material waits for the days you need a specific answer rather than the argument behind it.

Epilogue: Sovereignty as a Feature

This book opened on a hospital technology lead who could not use the best AI assistant in the world, because using it meant sending patient data somewhere she was not allowed to send it. I want to end there too, because her problem is the whole argument compressed into one person's frustration, and the way Mistral answers it is the thing worth carrying away when the model names and the prices in these pages have gone stale.

For most of the short history of modern AI, capability and control were sold as a package you could not unbundle. The most capable models lived on a handful of American companies' servers, reachable only by sending your data to them, on their terms. You could have the capability or you could keep control of your data, and for a great many people — the regulated, the cautious, the sovereign, the simply careful — that meant you could not really have the capability at all. The wall was invisible to the people it did not affect and absolute to the people it did.

Mistral's contribution, more than any single model or feature, is to treat that bundle as something you can take apart. By publishing the weights, by offering European infrastructure, by supporting self-hosting and private deployment, it turns "capability *or* control" into "capability *and* control" — and that conjunction, not any benchmark score, is the reason the company matters. The hospital lead does not need Mistral to be the smartest model on earth. She needs it to be capable enough *and* runnable on terms she is allowed to accept, and the second half of that sentence is the half nobody else was offering her.

That is what the title of this epilogue means. Sovereignty — control over where your data lives, who can touch it, which laws govern it, whether you can run the thing yourself — is not a compliance checkbox or a marketing word. It is a *feature*, in the most literal sense: a capability of the product that lets you do something you otherwise could not. For the people who need it, it is the most important feature there is, because it is the one that moves the tool from the "forbidden" column to the "usable" one. Everything else — the quality of Vibe, the size of the context window, the cleverness of the agents — is secondary to the simple fact that you are allowed to use it.

But the same honesty that ran through every chapter has to close the book too. Sovereignty as a feature comes with a price, and the price is real. The control Mistral hands you is also responsibility it hands you (Chapter 26). The openness is genuine but not unlimited, and it has shifted before (Chapters 14 and 25). The capability is frontier-class but not always frontier-leading; the ecosystem is thinner, the polish younger, the breadth narrower than the American giants' (Chapter 25). Choosing Mistral is choosing a set of trade-offs, and the trade-offs are favorable exactly when control, openness, and compliance are what you need, and unfavorable exactly when raw capability or ecosystem depth is the only thing that matters. The clear-eyed user knows which situation they are in, and plenty of organizations end up keeping both on hand, reaching for whichever one the task in front of them rewards.

Here, I think, is the part that lasts. The specific models in this book will be replaced; Large 3 and Medium 3.5 will give way to whatever comes next, the prices will move, Vibe will grow features and shed rough edges. But the idea Mistral put on the table — that you should not have to surrender control of your data to get access to capable AI, that openness and sovereignty can be features of a serious frontier product rather than the consolation prize of a weaker one — that idea does not expire. If anything it deepens, because as AI moves deeper into hospitals and banks and governments and the regulated core of the economy, the number of people who hit the hospital lead's wall only grows. A capable AI you are actually allowed to run is not a niche concern. It is, increasingly, the whole ballgame for the institutions that matter most.

So the closing advice is simple. Know which kind of user you are. If you need the absolute capability ceiling and the deepest ecosystem and nothing constrains where your data goes, there are excellent American assistants and you should use them without guilt. But if you are the hospital lead — if the wall is real for you, if control is not a preference but a requirement, if you have been told that capable AI is something you must watch other people use — then Mistral is the company that built the door, and walking through it is the point of everything in these pages. Capability you are allowed to control is not a lesser kind of capability. For the people who need it, it is the only kind that counts.

What follows is reference — the model details, the API quickstart, the self-hosting notes, the glossary, the resources — there for the day you need a specific answer rather than the argument behind it. The argument, by now, you have.

Appendix A: Model Reference

This is the most perishable page in the book, and I want to say so plainly before you read a single number. Model names, version strings, parameter counts, context sizes, license terms, token rates, and which model powers which surface are *exactly* the details that move on a scale of weeks to months at Mistral. What follows is a mid-2026 snapshot, offered so you have a shape to reason against — but **the only authoritative source is Mistral's own current documentation and model cards, and you should verify everything here against them before making any decision that turns on a figure.** Treat this appendix as "roughly the shape of things as of writing," not as current fact. The structure is durable; the numbers are not.

The Family at a Glance (Mid-2026)

Mistral's models as of this writing belong to the **Mistral 3** generation, and the two flagship-class members are the ones to know by name (Chapter 4):

Mistral Large 3 — the most capable model in the family. A large **sparse mixture-of-experts (MoE)** design, on the order of **675 billion total parameters** with only a fraction active per token, released under the permissive **Apache 2.0** license. It is the model to reach for when capability matters most, and its open license makes it the centerpiece of Mistral's sovereignty story: you can obtain the weights, run it yourself, and deploy it without anyone's permission (Chapter 14).

Mistral Medium 3.5 — a strong, efficient model and the engine behind much of **Vibe**. A **dense** model of roughly **128 billion parameters**, built from a merged lineage and released under a **modified-MIT license** (a permissive license with carve-outs for very-large-revenue commercial use — read it before you assume; Chapter 14). Medium 3.5 is the workhorse: most of what Vibe does in everyday use it does on a Medium-class model, because the capability-per-cost is excellent.

Smaller, faster open models. Below the flagships sits a range of smaller models — efficient enough to run on modest hardware, fast and cheap to serve, and well suited to high-volume or latency-sensitive work where the largest model would be overkill (Chapter 19). These are the models that make self-hosting practical on hardware most teams already have.

All of the above is verify-flagged. Names, parameter counts, and the exact roster shift with each release; **confirm the current family against Mistral's documentation**, because by the time you read this the version strings will likely have moved.

Context Window

Both flagship-class models offer a **large context window — on the order of 256,000 tokens** (Chapter 4): roughly a long book's worth of text held in mind at once. Long documents, large codebases, and extended sessions fit comfortably where they would have overflowed an earlier model. Treat the exact figure as a check-the-docs number — context sizes change and sometimes differ by model and by surface — but the durable point holds: this generation has room to spare.

Licenses: Open Weight, with Real Differences

The single most important thing this appendix can teach you is that **"open weight" is not one thing** (Chapter 14). Mistral publishes the trained weights of its models — but under different licenses with different permissions:

Apache 2.0 (as on Mistral Large 3) is genuinely permissive: download, run, modify, fine-tune, self-host, and deploy commercially, with minimal strings. This is about as open as a model license gets.

Modified-MIT (as on Mistral Medium 3.5) is permissive for the great majority of users but carries **carve-outs** — typically an exception requiring a commercial arrangement above a large revenue threshold. For most readers it behaves like a fully open license; for a large enterprise it may not. The difference matters enough that you should read the actual license for any model you intend to build on, not assume from the family.

The honest frame, repeated from Chapter 14: Mistral is *meaningfully more open than the American flagships, with real and shifting limits* — not "everything is free for everyone."

Open-Weight Availability

Because the weights are published, you can obtain the open models directly — typically as downloadable weights from a model hub such as **Hugging Face** — and run them on your own infrastructure (Appendix C, Chapter 15). This is the mechanism behind the whole sovereignty proposition: the model can live inside your network, with nothing transmitted to anyone. Which specific models are available as open weights, and under which license, is itself perishable — confirm the current availability on the model cards before you plan around it.

API Token Rates (Verify Before You Budget)

For programmatic use through **La Plateforme** (Appendix B, Chapter 18), models are billed per token — input being what you send, output being what the model generates back. As of this writing, **Mistral Large 3** is in the neighborhood of **$0.50 per million input tokens and $1.50 per million output tokens**. That ratio, output costing more than input, is standard across the industry, and these rates put Large 3 in the competitive range for a frontier model.

Verify both figures against Mistral's current API pricing before you budget against them. API rates move; they vary by model within the family; smaller models cost less; and API billing is **entirely separate from any Vibe subscription** (Chapter 21). The numbers here are a reference point, not a quote.

Vibe Consumer Tiers (Structure, Not Prices)

Vibe — the assistant (Chapter 3) — is sold through a ladder of consumer and organizational plans: **Free**, **Pro**, **Team**, and **Enterprise**. That structure is the durable part. The exact prices and what each plan unlocks are perishable, and rather than print figures that will age, I have kept them out of this appendix on purpose: **Chapter 5 owns the plan prices and contents**, and Mistral's live pricing page is the authority. The one structural point worth carrying here is the one Chapter 5 makes: **Vibe subscription billing and La Plateforme API billing are two different meters** — paying for a Vibe plan does not buy you API tokens, and vice versa.

Structure Is Durable; the Numbers Aren't

Use this appendix to understand the *structure* — a Mistral 3 family with a most-capable flagship (Large 3, Apache 2.0, sparse MoE) and an efficient workhorse (Medium 3.5, modified-MIT, dense, powering Vibe), smaller open models below them, a roughly 256k-token context window, open weights you can download and run yourself, API token rates billed separately from Vibe, and a Free / Pro / Team / Enterprise plan ladder. That structure is reasonably durable. Every specific number, name, and license clause in it is not. Before you choose a model, budget for API usage, pick a plan, or build a deployment around a license, confirm the current reality against Mistral's own documentation and model cards. The book's job is to tell you what to look for and how to think about it; the provider's job is to tell you today's figures, and only the provider's figures are authoritative.

Appendix B: API Quickstart

This is the shortest path from nothing to a working Mistral API call through **La Plateforme**, Mistral's developer platform. It is a starter, not a manual — Chapters 18 through 21 are the real treatment of the platform, model selection, and cost. Use this to get a first request returning, then go there. One standing caution: the exact endpoint names, parameter shapes, and SDK details are the kind of thing that drifts between releases, so **confirm every specific against Mistral's current API documentation before you build on it.** The example below is illustrative and may have moved by the time you read it.

Step 1 — Get a Key

API access is separate from any Vibe subscription (Chapter 18, Chapter 21). You authenticate programmatic calls with an **API key**, created from your La Plateforme account after you set up a developer account and billing. Two practical notes before you copy anything down:

Treat the key as a secret. It authenticates spending against your account. Keep it out of your source code and out of version control — read it from an environment variable or a secrets manager, never paste it into a file you will commit.

Know which meter you are on. La Plateforme bills per token (Appendix A, Chapter 21), and that meter is **entirely separate from a Vibe Free, Pro, Team, or Enterprise plan**. The API is the right meter for scripting and automation, but it is a different bill, so know which one you are running against before you wire up something that calls frequently.

Step 2 — Make a First Request

A minimal first call sends the model a message and reads back its reply. The shape is the chat-completions pattern the industry has converged on, which means if you have called another frontier model's API, this will look familiar. A small illustrative example:

```
# Illustrative only — verify the SDK name, model id, and base URL against curren
t Mistral docs
```

```
import os
from mistralai import Mistral

client = Mistral(
    api_key=os.environ["MISTRAL_API_KEY"],
)

response = client.chat.complete(
    model="mistral-large-latest",
    messages=[
        {"role": "user", "content": "In one sentence, what is Mistral?"},
    ],
)

print(response.choices[0].message.content)
```

Two things make this work. The **client** points at Mistral's platform — the API base is around `api.mistral.ai`, which the official SDK targets by default, but **confirm the base URL against the current docs**, because it is exactly the kind of detail that drifts. The **model string** names the family member you want (Appendix A); something like `mistral-large-latest` selects the current flagship, but verify the current model identifier, because version strings are the first thing to age. Everything else — reading the key from the environment, sending a list of messages, printing the reply — is the ordinary pattern. **Confirm the SDK name, the base URL, the model identifier, and the call signature against Mistral's live API docs**; any of them may have drifted. If you prefer not to use the official SDK, the API also speaks the common chat-completions shape over plain HTTP, so a familiar client library can often be pointed at it by changing the base URL.

Step 3 — Choose Your Model Deliberately

The fastest way to overspend on an API is to send every request to the largest model. Mistral's family spans a range (Appendix A, Chapter 19): **Mistral Large 3** for the hardest work, **Medium 3.5** for the efficient middle, and smaller models for high-volume or latency-sensitive jobs where the flagship would be overkill. The thing to carry here is not a model string, which will drift, but the **habit**: pick the smallest model that clears the bar for the job, and reserve the flagship for the requests that genuinely need it. That single discipline does more for your bill than any other (Chapter 21).

A second option worth knowing exists but lives elsewhere in the book: because Mistral publishes open weights, you are not limited to the hosted API at all — you can run the same models on your own infrastructure (Appendix C, Chapter 15). The API is the fastest start; self-hosting is the sovereign one.

Where to Go Next

This appendix gets you to a first response and a mental model. The depth lives in the book:

Chapter 18 — La Plateforme for authentication, the platform surface in full, models and parameters, and building beyond a single call.

Chapter 19 — Choosing a Model for matching the model to the job so you neither overpay for simple work nor under-serve hard work.

Chapter 20 — Fine-Tuning for adapting a model to your own data and domain through the platform.

Chapter 21 — Cost and Deployment for budgeting against token rates and for deciding between hosted API and self-hosted weights.

Get one call returning with the snippet above, then read those. The fastest way to understand the API is to have already made it say something back to you.

Appendix C: Self-Hosting Notes

This appendix is the practical companion to Chapter 15. It collects, in one place, what you actually need to know to run Mistral's open-weight models on your own infrastructure — where the weights come from, what runs them, what hardware it takes, and what to expect when you do. It is deliberately conceptual rather than a step-by-step install guide: the specific commands, framework versions, and hardware recommendations drift fast, so **verify every specific against the current model cards and the documentation of whatever inference framework you choose.** What does not drift is the *shape* of the task, and that is what this page is for.

Why Self-Host at All

The whole reason this option exists is the one Chapter 1 opened with: self-hosting is how the data never leaves. When you run the weights inside your own network, no prompt, no document, and no output is transmitted to Mistral or anyone else — the model runs on machines you control, under rules you set. For regulated industries, European institutions, and anyone for whom data residency is a requirement rather than a preference (Chapter 16, Chapter 23), that is the difference between "interesting" and "usable." Self-hosting is also the answer to cost control at scale and to avoiding lock-in to any single provider. It is the most demanding way to use Mistral and the most sovereign.

Where to Get the Weights

Because Mistral publishes open weights (Chapter 14, Appendix A), you obtain a model by downloading its trained parameters — typically from a model hub such as **Hugging Face**, where Mistral posts model cards alongside the files. The model card is the document to read first: it states the **license** (Apache 2.0, modified-MIT, or otherwise — and the license governs what you are permitted to do, so confirm it covers your use), the parameter count, the context window, and any notes on running the model. Confirm the current download location and license on the card itself, since both are exactly the kind of perishable detail this book refuses to pin down.

What Runs the Weights: Inference Frameworks

Raw weights are not a running service; you need an **inference framework** to load them and serve responses. The widely used open option is **vLLM**, which is built for serving large language models efficiently and exposes an API your applications can call — often in a shape compatible with the common chat-completions pattern, so code written against the hosted API can frequently be repointed at your own server with little change. Other frameworks exist and each has trade-offs around throughput, memory efficiency, and ease of setup. The durable point is the layering: **weights** (what you download) plus a **framework** (what serves them) equals a running model. Which framework, and which version, is a check-the-docs choice.

The Hardware Reality: It Scales with Size

Here is the trade-off that governs everything else: **hardware requirements scale with model size.** A model's parameters have to fit in memory to run well, and the largest models are large.

The smaller open models run on modest hardware — a single capable GPU, sometimes hardware a team already owns — which is what makes self-hosting practical for high-volume, latency-sensitive, or cost-sensitive work (Chapter 19).

The flagship-class models — a 675-billion-parameter Mistral Large 3, even as a sparse MoE that activates only a fraction of its parameters per token — demand serious GPU infrastructure: multiple high-memory accelerators, and real attention to memory, interconnect, and serving configuration. This is data-center or well-provisioned-cluster territory, not a workstation.

This is the single most important planning fact in the appendix, and it cuts both ways. It is why **model choice is a deployment decision, not just a capability one** (Chapter 19, Chapter 21): the right self-hosted model is often the *smallest* one that clears the bar for your task, because every step up the size ladder is a step up the hardware bill. Verify the current memory and GPU guidance on the model card before you provision anything — the numbers move with each release and with each advance in quantization and serving technique.

What to Expect

A few honest expectations, from having watched teams do this:

You own the operations. Self-hosting means you run the service — capacity, uptime, updates, and scaling are yours. That is the price of sovereignty, and for the organizations that need data residency it is a price worth paying. Mistral offers enterprise support and private-deployment arrangements for teams that want help carrying it (Chapter 17).

Quantization buys you headroom. Running a model at reduced numerical precision shrinks its memory footprint, often letting a larger model fit on smaller hardware with modest quality cost. It is one of the main levers for making a big model affordable to host, and the model card or framework docs will say what is supported.

The smaller models are more capable than you expect. For a great many real tasks, a smaller self-hosted model is entirely sufficient, and choosing it well (Chapter 19) is most of running a sensible self-hosted deployment.

Start with the hosted API to learn, then self-host to control. The quickest way to understand a model's behavior is through La Plateforme (Appendix B); the quickest way to satisfy a data-residency requirement is to host the weights yourself. Many teams do both — prototype on the API, deploy on their own GPUs.

For the full treatment of running models yourself — the case for it, the deployment spectrum, and how it fits the sovereignty story — see **Chapter 15**, with **Chapter 17** on private enterprise deployment and **Chapter 21** on weighing hosted cost against self-hosted cost. This appendix is the field-notes version; the chapters are the argument.

Appendix D: Glossary

Terms as this book uses them. Where a term has a broader industry meaning, the definition here is the sense relevant to working with Mistral.

Apache 2.0. A permissive open-source software license, used for some Mistral model weights — including, as of this writing, Mistral Large 3. It allows you to download, run, modify, fine-tune, self-host, and deploy commercially with minimal restrictions. About as open as a model license gets. See Chapter 14, Appendix A.

Connectors. Integrations that let Vibe reach into your other apps and tools — so the assistant can act across your connected services rather than only within the chat box. The plumbing behind Work mode's ability to delegate multi-step tasks. See Chapter 8.

Context window. How much text a model can hold in mind at once in a single conversation — your prompt, any documents you have given it, and everything said so far, counted together. When you exceed it, the earliest material falls out of view. Mistral's flagship-class models offer a large window, on the order of 256,000 tokens. See Chapter 4, Appendix A.

Data residency. The requirement that data be stored and processed in a particular jurisdiction. Mistral's European infrastructure and its self-hosting option exist largely to satisfy data-residency rules that a US-based provider structurally cannot meet. The practical core of the sovereignty proposition. See Chapter 16, Chapter 23.

Fine-tuning. Adapting a base model to your own data and domain by training it further on examples you supply, producing a model that behaves the way your task needs. Available through La Plateforme and, with open weights, on your own infrastructure. See Chapter 20.

La Plateforme. Mistral's API and developer platform — where you get an API key, call the models programmatically, fine-tune, and build applications. Billed per token, separately from any Vibe subscription. See Chapter 18, Appendix B.

Le Chat. The original name of Mistral's assistant, before it was renamed **Vibe** in 2026 as it grew from a chatbot into a work agent. If you have heard of Le Chat, Vibe is what it became. See Chapter 3.

Libraries. Vibe's feature for saving and organizing reusable material — the things you return to — so they are at hand across conversations rather than re-pasted each time. See Chapter 8.

Memory. Vibe's ability to retain useful context about you and your work across conversations, so you need not re-establish it each session. See Chapter 8.

Mistral. Europe's flagship AI company — French, founded in 2023 — and the subject of this book. It builds both the Vibe assistant and a family of open-weight models, with openness and sovereignty as its defining traits. See Chapter 1, Chapter 2.

Mistral Large 3. The most capable model in the Mistral 3 family as of this writing — a large sparse mixture-of-experts model, on the order of 675 billion parameters, released under the permissive Apache 2.0 license. The model to reach for when capability matters most. See Chapter 4, Appendix A.

Mistral Medium 3.5. A strong, efficient dense model of roughly 128 billion parameters, released under a modified-MIT license, and the engine behind much of Vibe. The workhorse of the family. See Chapter 4, Appendix A.

modified-MIT license. A permissive license, derived from the MIT license, used for some Mistral model weights — including Mistral Medium 3.5. Permissive for most users, but with carve-outs (typically a commercial-arrangement requirement above a large revenue threshold), so read the actual terms for any model you build on. See Chapter 14, Appendix A.

MoE (mixture of experts). A model architecture in which only a fraction of the model's parameters — a subset of specialized "experts" — activate for any given token, so a very large model can run far more efficiently than its total parameter count suggests. Mistral Large 3 is a sparse MoE. See Chapter 4.

open source. The full ideal in which not only a model's weights but its training code, data, and process are openly available under an open license. A broader and stricter standard than "open weight." Mistral is open *weight* and, for some models, under genuinely open *licenses*, but the two ideas are not the same. See Chapter 14.

open weight. A model whose trained parameters (the weights) are published for download, so you can run, inspect, fine-tune, and self-host it — as opposed to a closed model you can only reach over someone else's API. The heart of what makes Mistral different. Not identical to "open source": the weights are open, the license governs what you may do, and the licenses vary. See Chapter 14.

self-hosting. Running a model's open weights on infrastructure you control — your own hardware or private cloud — so no data is transmitted to any provider. The most sovereign way to use Mistral; hardware requirements scale with model size. See Chapter 15, Appendix C.

sovereignty. The principle that an organization should control where its AI runs, where its data goes, and under whose laws — rather than being forced to hand sensitive information to a foreign black box. Mistral's reason to exist, achieved through open weights, European infrastructure, and self-hosting. See Chapter 1, Chapter 23.

token. The unit models read and write in — a chunk of text, smaller than a word on average — and the unit API usage is billed and context is measured in. A 256,000-token context window is roughly a long book's worth of text. See Chapter 4, Chapter 21.

Vibe. Mistral's frontier AI assistant — formerly Le Chat — organized around three modes (Work, Code, Chat) and sold through Free, Pro, Team, and Enterprise plans. The friendly face of Mistral; the open models are the substance beneath it. See Chapter 3, Chapter 6.

Vibe Chat. Vibe's mode for the classic conversational experience — asking, writing, and reasoning in a back-and-forth. The mode most like a traditional chatbot. See Chapter 6.

Vibe Code. Vibe's mode for software development — writing, explaining, and working with code. See Chapter 6.

Vibe Work. Vibe's mode for delegating multi-step tasks across your connected apps — the agent surface, where Vibe acts rather than only answers. See Chapter 6.

Appendix E: Resources and Further Reading

This book is a snapshot of a tool and a company that move fast. Models ship, licenses shift, prices change, and the lineup is renamed on a cadence of weeks to months. So the single most important resource is whatever is *current*. This appendix points you at where to find current truth and how to keep up given how quickly the specifics drift. When this book and an official source disagree, the official source wins — it is versioned with the tool; this book is not.

The Authoritative Sources

Mistral's official site and product documentation. The authoritative reference for what Vibe can do, which modes and features exist, how each works, and which plans contain what. This is where you confirm anything specific — a capability, a plan's contents, a feature's availability — and where you will find what changed since this book was written. Reach it from Mistral's main site rather than any link I print here, because URLs are exactly the kind of fragile detail that breaks.

La Plateforme API documentation. The only authoritative source for the developer surface — the API base, endpoints, model identifiers, parameters, authentication, fine-tuning, and rate limits (Chapters 18 through 21, Appendix B). For anything you intend to build on, the live API docs beat any written reference, including this one. Find them through your La Plateforme account.

Model cards (on Hugging Face and Mistral's site). The only authoritative source for the open-weight models themselves — the current roster, parameter counts, context sizes, **licenses** (Apache 2.0, modified-MIT, or otherwise), and the notes you need to download and run a model yourself (Chapter 14, Chapter 15, Appendices A and C). Read the model card before you self-host or build on any model, because the license governs what you are permitted to do and it varies by model.

Mistral's pricing pages. The only authoritative source for Vibe plan prices and contents and for API token rates — all of which this book deliberately treats as perishable (Chapter 5, Chapter 21, Appendix A). Check here before any decision that turns on a number, and remember the two meters: Vibe subscriptions and La Plateforme API billing are separate.

Model announcements and release communications. For knowing what is new, what changed, and whether something you read about here has since moved. A company that ships and renames this often rewards keeping half an eye on its announcements.

Compliance and Regulatory Resources

Much of what makes Mistral matter is regulatory (Chapter 16, Chapter 23), and that landscape has its own primary sources worth going to directly rather than relying on any vendor's summary:

The EU AI Act and its official guidance. The governing regulation for AI in the European Union, and the backdrop for much of Mistral's sovereignty and compliance story. Consult the official EU sources for the current text and implementation timelines rather than secondhand accounts, since the obligations phase in over time and the details matter.

Your own jurisdiction's data-protection regime — GDPR in the EU, and the sector-specific rules that govern your industry. Whether Mistral clears your bar is ultimately a question your compliance function answers against these sources, not one this book can answer for you. The book explains why the sovereignty option exists; your regulators define what you actually need.

I have kept these deliberately generic and URL-free. Compliance resources are exactly where a fragile printed link does the most damage, and the official bodies are easy to reach by name.

Companion Books from Finnoybu Press

Mistral sits in a landscape of frontier assistants and AI tools, and several Finnoybu Press titles cover the neighboring ground. If Mistral is one piece of your toolkit, these cover the others:

On the assistants Mistral competes with (Chapter 24's comparison):

- The product guides to ChatGPT, Claude, Gemini, and Grok — the deeper treatments of the US flagships Mistral stands against, for readers weighing which to reach for when, or running more than one alongside Mistral.

On the broader skills that working with AI rewards:

- *AI Prompt Engineering* — the craft of instructing models well, which underlies getting good work out of Mistral across every chapter.
- *AI for Everyone* and the wider product-guide catalog — for rounding out how AI fits into the rest of your work beyond any single assistant.

See **press.finnoybu.org** for the current catalog and the rest of the Extended Series.

A Closing Word on Staying Current

The half-life of specific facts about Mistral is short, and that is not a flaw in the company or this book — it is the cost of writing about a field moving this fast. The durable parts of this book — what openness and sovereignty are *for*, how to use Vibe's three modes well, when to self-host and when to use the API, how to think about licenses, data residency, cost, and responsible use — will serve you across many versions. The perishable parts — exact model names, parameter counts, prices, plan contents, context sizes, license clauses, API specifics — you refresh from the sources above whenever they matter. Hold the durable, look up the perishable, and you will stay current without needing a new book every quarter.

www.ingramcontent.com/pod-product-compliance
Lightning Source LLC
LaVergne TN
LVHW060033210926
848481LV00013B/3849
9798950533235